D1260576

The
Pleasure
Instinct

The Pleasure Instinct

Why We Crave Adventure, Chocolate, Pheromones, and Music

Gene Wallenstein, Ph.D.

WILEY

John Wiley & Sons, Inc.

Published by John Wiley & Sons, Inc., Hoboken, New Jersey
Published simultaneously in Canada

For general information about our other products and services, please contact our Customer Care Department within the United States at (800) 762-2974, outside the United States at (317) 572-3993 or fax (317) 572-4002.

Wiley also publishes its books in a variety of electronic formats. Some content that appears in print may not be available in electronic books. For more information about Wiley products, visit our web site at www.wiley.com.

Library of Congress Cataloging-in-Publication Data:

Wallenstein, Gene, date.
    The pleasure instinct : why we crave adventure, chocolate, pheromones, and music / Gene Wallenstein.
      p. cm.
    Includes bibliographical references and index.
    ISBN 978-0-471-61915-4 (cloth)
      1. Pleasure. I. Title.
  BF515.W29   2008
  152.4'2—dc22

                                              2008041537

Printed in the United States of America

10 9 8 7 6 5 4 3 2 1

For Kai, Ren, and Finn, of course

Contents

Acknowledgments

Whatever good the reader finds in this book can be traced to those who have, in one way or another, taught me over the past couple of decades about human nature. To those who have inspired me (and countless others) through the years with their lectures, conversations, writings, and generous time, I thank John Allman, David Barash, Kent Berridge, T. Berry Brazelton, David Buss, Michael Cabanac, Michael Crawford, Richard Davidson, Richard Dawkins, Terrence Deacon, Irven DeVore, Jared Diamond, Ellen Dissanayake, Robin Dunbar, Paul Ekman, Howard Eichenbaum, Nancy Etcoff, Steven Gangestad, Fred Gage, Elizabeth Gould, Steven Jay Gould, William Greenough, Dean Hamer, William Hamilton, Michael Hasselmo, Marc Hauser, Dee Higley, Sarah Blaffer Hrdy, Nicholas Humphrey, Thomas Insel, Victor Johnston, Jerome Kagan, J.A. Scott Kelso, Ray Kesner, Melvin Konner, Judith Langlois, Joseph LeDoux, Paul MacLean, John Manning, Andrew Meltzoff, Michael Merzenich, Geoffrey Miller, Steven Mithen, Sheri Mizumori, Anders Moller, Allan Nash, Ulric Neisser, Charles Nemeroff, Jaak Panksepp, Steven Pinker, Mark Ridley, Terry Robinson, Norman Rosenthal, Michael Ryan, Robert Sapolsky, Ellen Ruppel Shell, Devandra Singh, Georg Striedter, Donald Symons, Randy Thornhill, Sandra Trehub, Robert Trivers, Leslie Ungerleider, Ann Wallenstein, Greg Wallenstein, Claus Wedekind, George Williams, E. O. Wilson, Roy Wise, Amotz Zahavi, and Robert Zatorre.

I also wish to thank my editor at John Wiley & Sons, Christel Winkler; my literary agent, Jim Hornfischer; and Tom Miller at

Wiley for their invaluable encouragement and steadfast commitment to the project.

My dear wife, Melissa, is owed a level of gratitude that is impossible to repay. Her love, companionship, support, and intellectual stimulation have been at the core of my life for the past fifteen years and have provided continual inspiration during the writing of this book. Finally, for teaching me what is most profound about human nature, I wish to thank my beautiful children, to whom this book is dedicated.

Part One

The Pleasure Instinct and Brain Development

Chapter 1

Foibles and Follies

If you prick us, do we not bleed? If you tickle us, do we not laugh? If you poison us, do we not die? And if you wrong us, shall we not revenge?

—William Shakespeare, *The Merchant of Venice*

Human nature exists.

—Melvin Konner, *The Tangled Wing*

W hy does pleasure exist?

Beyond academic circles one seldom hears this question. In daily life, as we move through the minutiae of meetings, ready the kids for school, manage a household, and take care of the basic necessities, we're more likely to seek new ways to pursue pleasure than ponder its existence. Pleasure, like fear and fire, is a natural force that humans have sought to harness and control since their beginnings. The pleasure instinct—evolution's ancient tool for prodding us in the directions that maximize our reproductive success—has created a staggering panorama of behaviors, pathologies, and cultural idioms in our modern lives that often bewilder and beguile.

This book is a biography of sorts, a chronicle of the relationship between humans and pleasure. As the story is told, we will address some of the deepest questions that have been asked about human nature through recorded history and undoubtedly beyond. To understand pleasure, we must know its history and evolution. How is it that the human mind experiences pleasure in mere shapes and colors, textures and touch, myths and stories? Why does humor relieve tension? Why does music invigorate us—to dance, swoon, make love, or march off to battle—while many other noises leave no mark? Why do social attachments make us feel good? Do other animals experience pleasure? Why do we find babies so darn cute? And how is it possible that pleasurable feelings can be elicited from such an astonishingly wide array of events ranging from the mother's gaze at her newborn to the addict's anticipation of his next high?

Philosophers and spiritual leaders have debated the value and nature of pleasure for centuries, often comparing it to its more abiding sibling, happiness. The two are related, of course, but most of us, from saint to sinner, have never doubted which of the pair would make the best honeymoon companion. Happiness is often said to be a "gift for making the most of life" or "enjoying the simple things." Pleasure is a hedonistic reflex, a burning impulse to abandon rational thought altogether and immerse oneself in the moment. Happiness is an abstraction, constructed from our social and moral identities—a carefree stroll on the beach, 2.3 children and a white picket fence, a sense of accomplishment. The pleasure instinct, like the survival instinct, is pure biological imperative fueled by an ephemeral reward so fevered and beautiful with desire that it can drive us to extraordinary lengths. Happiness is a Norman Rockwell painting hanging over your fireplace on a cold winter's eve. Pleasure is the warmth and aesthetic beauty of the flames, the heat beating on your skin.

Pleasure is experienced in a multitude of colorful ways—the ecstasy of a sexual encounter, the epicurean delight of chocolate, the delivery of a punch line. Yet despite this it has a central core of universal features that cuts across all human cultures and historic periods. In this respect, we are all deeply connected by both the gifts and constraints that natural selection and adaptation have afforded us.

We live in an antidepression era, dominated by a seemingly insatiable appetite for happiness, and it is critically important to our individual and societal health (and happiness) to understand why this is the case. We can't get no satisfaction. We go on spiritual quests, read all the right books, join health spas, travel, buy new cars, eat out, watch cable TV. We all want a piece of it—bliss, elation, cheer, the primrose path, spice, titillation, glee, exuberance, mirth, joy, and jubilation. How about vice, addiction, lust, malfeasance, adultery, a monkey on your back, obsession, and perversion? Our modern brains, forged from the grist of evolution's mill during our stay as hunter-gatherers, must deal with contemporary conditions that are radically different from—and in some cases in direct opposition to—the ancestral environments in which more than 97 percent of our history has been lived. Thus the importance of understanding why pleasure evolved, how the context and selection factors that shaped its evolution differ vastly from the environmental circumstances we face today, and the personal and societal consequences of these differences cannot be overstated. And certainly not ignored.

Pleasure is not an epiphenomenon, a lucky happenstance of neurons being in the right place and firing at the right time. It has evolved to serve a very specific and adaptive set of functions from our distant past. The genes that encourage the expression and feeling of pleasure are success stories of natural selection—they are still around. Therefore, in our quest to understand the psychological, biological, and cultural foundations of pleasure in the modern world, we must consider what problems pleasure solved for our ancestors. If the pleasures did not provide a functional solution to some selection factors faced by our earlier brethren, the genes that shape their expression and feeling would be long gone, into the dustbin of ecological time like most others.

Darwin without "Social Darwinism"

Understandably, some people twinge when Darwinism and human nature are mentioned in the same breath—Darwin himself made virtually no reference to humans in his great work *The Origin of Species*.

Our fear, perhaps, lies in what we foresee as a sterile, eugenic existence such as that thrust upon the characters in Aldous Huxley's *Brave New World*. Most of us genuinely resent the notion that our behaviors, thoughts, and feelings—the very ingredients that make humans extraordinary—are shaped, even in part, by biological and genetic factors. We refuse to accept that our genes chain us to a destiny preordained by proteins. But, as we'll see throughout this book, nothing in modern Darwinian theory claims this to be the case. Indeed, understanding why emotions evolved, particularly the pleasure instinct, can have a profound and positive impact on daily life by showing readers how pleasure influences the way we make aesthetic, social, and moral choices, and learn from our mistakes.

Until recently, scientists have concentrated almost exclusively on studying how social and cultural factors shape emotions. Academics have typically shied away from using evolutionary principles to study slippery subjects such as pleasure because they often fear that such an approach leaves little or no room for the role of experience. We've heard it so many times: "People like things because they learn to like them, not because their genes tell them to." We're not programmed with innate preferences, the argument goes; we learn what is pleasurable through trial and error. And this is often the case. Yet studies have shown that rather general biases are present immediately after birth. For example, newborns prefer the taste of sweets to sours; a smile to an expressionless face; symmetrical to asymmetrical objects and scenes; and rhythmic to random sounds. These preferences emerge long before the infant ever encounters a cookie or hears its first joke. This book explores the many innate proclivities that have evolved in humans as a result of our pleasure instinct, and examines how they dramatically shape our brains, behaviors, thoughts, and feelings.

In the last twenty years a scientific revolution has been under way that marks a significant departure from how human behavior has been studied in the past. The new thinking rests on the firm belief that to make sense of human nature, one must consider how the human mind was molded by both natural and sexual selection. Instead of simply asking, How does the mind solve problem X?

a better question is, Why was that particular mind/brain mechanism selected for during our earlier history as hunter-gatherers?

Some claim this revolution began with the 1975 publication of E. O. Wilson's now classic book *Sociobiology*, which examined the way selection factors influence reproduction strategies. Rarely has a scientific text produced such a strong sense of political outrage, particularly among social critics who saw it as a scientific justification for patriarchal societies. Posters were placed all over Harvard University inviting students to attend and disrupt Wilson's classes with noisemakers, and he was attacked in the press by many of his colleagues.

Yet the revolution has continued to gather steam, finding support and a more solid foundation from such diverse fields as molecular biology, behavioral genetics, cognitive psychology, cultural anthropology, and neuroscience. Sociobiology and its heir apparent, evolutionary psychology, have now grown at an exponential rate, with thousands of researchers using them as the main paradigms through which they investigate human behavior and indeed culture.

For our purposes, we will use both evolutionary and developmental principles to help us navigate the rambling terrain of pleasure—from its ancient landmasses that gave rise to modern landscapes, to its largely unexplored hinterlands. Pleasure, as we shall see, is the "common currency" that regulates the way humans self-stimulate their own brain growth and maturation. Human babies, for instance, are exceedingly discriminating in what they prefer to look at, listen to, feel, taste, and smell. These innate biases—from a love of primary colors, to a fondness for prosody—ensure that infants seek out the best kinds of sensory experiences for promoting normal brain development in their early years of life. Such biases persist well beyond the critical periods that every parent is familiar with, and form a system of positive reinforcers that profoundly impact adult cognition and behavior. In the chapters that follow we will address long-standing questions and learn why pleasure is ultimately a regulator of development. The issues that arise are at the very core of what it means to be human, and give us a glimpse of what we can reasonably expect of human nature. Why did pleasure evolve?

What are the evolutionary advantages, biological realities, and consequences of pleasure?

- How does pleasure fine-tune the brain? Why are certain sensory
  experiences more pleasurable than others?
- What evolutionary/developmental factors govern our attraction
  and attachment to friends, lovers, relatives, and offspring?
- What makes sinning so much fun? How did addictive behaviors
  evolve and what can natural selection and developmental principles tell us about treatment?
- Why is laughter contagious? How is it related to aggression?
  Are mice ticklish? Do animals experience love and joy? Why do
  we smile when pleased?
- Is there such a thing as a universal set of aesthetics? Why do
  some of us see art while others see only squiggly lines? What is
  it that makes some people more physically attractive than others? Why do we like the sounds of wind, thunder, and flowing
  streams? Why do we find certain environmental landscapes so
  aesthetically pleasing?
- Why do so many of us take pleasure in thrills and chills—from
  parachuting out of airplanes to riding roller coasters to watching horror flicks? How are phobias related to thrill-seeking?
- And perhaps most important of all—how can we use what science is now learning about the pleasure instinct to improve our
  quality of life?

This is but a small sampling of the questions that arise naturally
when we ask: Why does pleasure exist? Fundamental answers to
these questions will not be found by generating a hodgepodge of
disjointed theories, each tied to a particular issue. Instead, when
asked against the backdrop of evolution, they reveal the framework
of a new worldview that is beginning to change the way we think
about human nature. The story of how the pleasure instinct evolved
and continues to function today begins with our first steps into the
cognitive niche.

The next two chapters of this book are dedicated to exploring these initial steps into the cognitive niche and provide a conceptual foundation for understanding the role of pleasure in the evolution of our species. Chapters 4 through 8 detail how the pleasure instinct facilitates normal brain growth and development in each of the five primary senses—touch, taste, smell, audition, and vision. Chapters 9 through 11 provide three examples of how the pleasure instinct impacts our everyday lives, including how we choose mates and why we love rhythm so much, and provides a new perspective on addictive behaviors. Finally, chapter 12 summarizes this material and considers the open questions that await answers from future research.

Chapter 2

How to Win Friends and Influence People

"I think therefore I am" is the statement of an intellectual who underrates toothaches.

—Milan Kundera, *Immortality*

Homo sapiens . . . can rightfully be called the babbling ape.

—Edward O. Wilson, *Consilience*

In most families there is nothing more exciting than the appearance of a new baby. In ours, the latest addition is my little niece Kathleen, who now in her fourteenth month can do so many amazing things, most of which we take for granted but are really miracles of development. A few months ago she joined the ranks of fellow bipeds and meanders about the house awkwardly, resembling a slightly inebriated little sailor making her way home after last call. She can recognize objects as being unique and distinct from others, no longer labeling everything uniformly as "daht." And she has an amazingly complex palate of emotional expressions, the full range of which, I have come to realize, can be displayed with little or no notice. But this is just the beginning.

It's been clear for many weeks that Kathleen can understand far more than she is able to verbalize. Sitting next to her at dinner the other night, I noticed she was trying to catch a balloon tied to the back of her chair. "Do you want me to get that for you, sweetie?" I asked. And then, all of a sudden, it happened—she said, "Yeah . . ." At last, contact! There was a real person inside that little body. Our brief exchange didn't grab the interest of those around us, but I was astonished by the unexpected exactness of her answer. For the first time, I truly felt we had made a connection.

Later that evening while talking with a friend about my dinner conversation, I tried to explain why I was so taken aback by my niece. Surely it is to be expected that she'll begin to talk sooner or later, but my surprise arose from two levels of awareness. On the first level, it's staggering to think of the mechanistic and computational achievement it is to extract meaning from mere acoustic energy—sound waves thrust in your direction from the peculiar manner in which people modulate their breath as they exhale. The biological and psychological capacities that support the many processes in between auditory sensation and language interpretation can (and do) fill volumes in university libraries. Then there is the other side, language production. After Kathleen has interpreted my question and decided on an answer, she must mold her young articulators into the correct spatial arrangement, which varies over time, to create the proper sound waves that will have meaning for me, the listener. And finally, there is all that fancy neural processing in between language interpretation and production that is made evident when we realize that Kathleen's answer does not result from some simple stimulus-response pairing, a monosynaptic reflex, or a bit of classical conditioning. Rather, a conscious, self-referential decision was made.

In his book *The Language Instinct*, cognitive scientist Steven Pinker marvels at this peculiar trick humans have evolved for communication:

As you are reading these words, you are taking part in one of the wonders of the natural world. For you and I belong to a species with a remarkable ability: we can shape events in each other's brains with exquisite precision. . . . Simply by making noises with our mouths, we can reliably cause precise new combinations of ideas to arise in each other's minds.

He makes the important point that the miracle of language is not just in its mechanics—sound waves bouncing off a cochlear, larynx, and pharyngeal openings constricted just so—but in the functional properties that emerge with its usage, namely the exchange of information that may come in any number of forms, such as those relating to nature, technology, social identity, physical health, emotions, and so on.

This brings me to the second reason I was so surprised by Kathleen. When she answered, "yeah," I was impressed by her response on an intellectual level, yet at the same time, I felt an inexplicably strong emotional reaction, an attachment that formed instantly with this single syllable. I had heard her say words before, so it was not the mere occurrence of recognizable speech, but rather the context of the social connection that bound us that was so engaging. Just as the emergence of language has been shaped, in both the species and the individual, by the competitive forces of natural selection, so too has the appearance of emotions such as pleasure. The manner in which pleasure drives our biological need for social attachment and communication is the subject of this chapter, and it is an amazing story.

Precocious Primates

Ask an archaeologist what factors gave *Homo sapiens* the competitive edge over our neighbors—*Homo erectus* in Asia and Neanderthals in Europe—and they will likely describe the impressive transition from Oldowan stone tool use found at sites dating 2 million to 1.5 million years ago to the more sophisticated Levallois flake technology for making sharp blades. They will further comment on the explosion in variety and specificity of tools for different functions that appear in the archaeological record: elegant wood-carved spears used for hunting game; blades shaped into projectile points, end scrapers, chisels, and burins, all custom-made to match a particular task; and tools born from bone such as awls and needles. Ask anthropologists the same question and they will use the same archaeological data to remind us that earlier hominids tended to segregate their daily lives into different locations according to task. Tools were constructed in one location, food preparation in another, and so forth. *Homo sapiens*, on the other hand, are believed to have used a centralized location where all of these activities were performed together, providing an integrative and highly social aspect to everyday life.

But we are left with this daunting question: Why did this shift in social behavior occur? Put simply, why did our ancestors enter the "cognitive niche"? The survival of typical *Homo sapiens* depended critically on the possession of very basic factual knowledge and skilled techniques for managing their place in the habitat. They had to be able to locate food and know how to extract and prepare it for consumption. They had to learn where their predators were and how to avoid or defend against them. They needed to be familiar with the terrain and, at the very least, possess rudimentary navigation skills. The list goes on and on—just for basic survival. Such increasingly complicated knowledge can most effectively be learned in the context of a social community. In the words of British psychologist Nicholas Humphrey, such a community "provides both a medium for the cultural transmission of information and a protective environment in which individual learning can occur." In this sense, the primary role of intelligence in higher primates is not to produce great works of art or advance scientific achievement, but simply to hold society together.

Once a species begins on the path toward socialization, it is as if they were thrown on an evolutionary treadmill, and there is no going back. The emergence of social interactions ultimately leads to ever-increasingly complex social behaviors, social emotions, and group conduct that in turn develop a need for yet more complex social skills. This process is known in evolutionary biology as a "ratchet effect," somewhat akin to a gear that is only capable of moving in a single direction. The limiting factor, of course, is determined by the extent to which the adaptive consequences of social behavior outweigh its burden and eventual cost on successful reproduction. Those individuals who place too great an emphasis on socialization while neglecting other subsistence factors will have a reduced chance of survival into reproductive age and attracting a suitable mate.

Social interaction brings with it enormous potential for changing the way individual members of a group go about their day-to-day survival. Though basic subsistence is always challenging, life in complex societies such as those constructed by many primates is demanding in a very different way. There are clear benefits to be had for those members of the group who can manipulate the social structure of the clan by outmaneuvering their peers. Individuals must be adroit at reading the social cues of the group; predicting the consequences of their own behavior and that of others; and tallying the complicated balance sheet of advantages and losses that revolve around these myriad social transactions. Hence, social primates are required to be calculating beings by the very nature of the system they create and maintain. In such a system, social skill, communication, and intellect are inseparable.

The selection pressures that led to early hominids' growing need for more sophisticated subsistence technology contributed to two important changes in their social behavior: (1) they granted offspring a longer grace period of dependence on adults, free to learn about their habitat through play, exploration, and experimentation; and (2) they encouraged greater interaction across generations whereby the young learn about subsistence technologies from elder, more experienced teachers. These shifts markedly widened the age range of the communal setting, and brought the very young into contact with the very old, resulting in particularly difficult social challenges. Both the older and the younger members of a community tend to be most dependent on the core adults of the group; thus an evolutionary mechanism must exist to facilitate or encourage the adults to cater to the whims, desires, and needs of these two groups. There must be some adaptive benefit for the adults that outweighs the cost to them in caring for the young, old, sick, and infirm. That adaptive benefit, of course, is mediated by the pleasure derived from social bonding—the pleasure I found in the exchange with my niece.

Many scholars agree that two behaviors probably provided the survival edge that benefited *Homo sapiens* over their contemporaries: the evolution of social attachment and language. I believe that both social attachment and language evolved from selection factors I call proto-emotions. These are basic, instinctual emotions that are exhibited by many primates: pleasure, fear, anger, disgust, sadness, surprise, and the various hungers. Proto-emotions have a very quick onset and are short-lived, almost like a reflex. The social emotions—for instance, happiness, maternal love, sexual love, infatuation, pride, and admiration—differ from proto-emotions in that they consist of long-lasting behavioral and mood states that typically outlast precipitating conditions and increasingly depend on a capacity for self-reference and reflection. These are the modern emotions we experience today that evolved from conditions in our ancestral past, remnants of a prior age where life was lived as hunter-gatherers on the open plains of the savanna. In the pages that follow, we'll explore how pleasure led to the evolution of social attachment and language, and most importantly, how it shaped the positive social emotions that reverberate through our lives so profoundly today. Why did our instinct for pleasure drive us to become such loquacious, social creatures? And how did this newfound love of gabbing, gossip, and group affiliation result in modern emotions such as love, lust, happiness, and joy?

The Language Link

We take for granted that language can illuminate what is subjective, the amorphous yet innumerable feelings, thoughts, and inklings that mix through our minds like hot and cold currents every moment of the day. We like to think that other humans share this dizzying internal menagerie, or at least some parts of it. But what would a dog say if it could speak a human language? Would a dog's inner emotional experiences be close enough to a human's so that a common lexicon might emerge? Could we really learn more about the thoughts and feelings of animals—what is in their hearts and minds—if we could decode their vocalizations?

The linguistic philosopher Ludwig Wittgenstein argued that all truths, be they emotional, moral, aesthetic, or intellectual, are known only through experience. He suggested that they lose their real value and meaning in the telling, and that language is merely a form of depiction, a representational system that inevitably fails to characterize our genuine nature since it can only work through analogy. Thus, even if we successfully decode an animal's sounds, we could not truly understand them because language is but a mirror of reality rather than the genuine object, and an animal's reality, as the argument goes, is too far removed from our own. Language is the finger pointing at the moon, not the moon itself. In cognitive science circles, this is known as the representation problem—deciphering how symbols, such as language, map onto subjective experiences, such as feelings and thoughts. The representation problem, of course, extends to all symbol-using species, and we will revisit it throughout this book.

A kissing cousin to this line of argument is the classic linguistic problem of induction—how one infers the referent of a word from a speaker's vocalizations and behavior. Imagine you are a linguist encountering a newly discovered human population. One of the clan members shouts "agovi" as a turtle saunters by. Your first guess, probably, is that *agovi* means "turtle." This is a perfectly reasonable inference, since at least in English, comments elicited by an object typically refer to the object itself. But this is premature because agovi may also refer to animals or objects that move slowly, have shells, are hard and spherically shaped, are smaller than a house but larger than a breadbox, or are the most important ingredient for soup. The induction problem shows that any attempt to determine word meaning strictly from behavior is in deep trouble, since there are simply too many possible interpretations for any specific action. How did the first language emerge, then, if we can't even get past single words? I believe a reasonable alternative is that hominids' initial foray into semantics, and perhaps the genesis of structured language, were driven not by their desire to label everything in sight, but rather by their common need to exchange emotional information.

How did proto-emotions, particularly pleasure, foster the evolution of modern language? There are many theories that are fun to consider. Did language develop in response to the increasingly complicated social lives our ancestors lived, or is it perhaps the other way around, a new tool that evolved for other reasons—a spandrel—that facilitated greater socialization? What did the first languages sound like and how did they facilitate social attachment and the development of modern emotions?

Although it's impossible to jump in a time machine and listen in on a prehistoric town meeting, we can adopt a different strategy. We can look to a source of information that will help us decipher how the pleasure instinct may have shaped social attachment and linguistic life for hominids—the emergence of spoken language in children. This is not an attempt to resuscitate the old Haekelian idea that ontogeny (the development of an individual) recapitulates phylogeny (the development of a species). This notion is based on the assumption that the ontogenetic form being considered develops through a series of stages that are essentially re-creations of the adult forms of its evolutionary predecessors. We will employ a more modern view that has emerged recently, which instead emphasizes studying the embryological and developmental commonalities and differences among genetically similar species. This theoretical approach never arose in Haekel's day because it depends on the modern science of genetics. We will use it here because learning how an infant becomes linguistic can tell us a great deal about how language arose in our species as an important tool for emotional expression. "The human race began to talk as babies begin to talk," noted the psychologist Carl Johnston, ". . . in the prattle of every baby, we have a repetition in a minor key of the voice of the earliest man . . . by watching the first movements of speech in a baby, we see once more the first steps in articulate language, which the whole world of man took in dim ages long ago."

The Trouble with Tribbles

In ninth grade I remember reading a novel set in the distant future where a highly sophisticated artificial intelligence program is implanted into robots, eventually giving them enough brainpower to take over the world from humans, outthinking our every move. If you want a more probable scenario, one that at least makes sense from an evolutionary perspective, our mechanized subjugators are most likely to be adorably cute little creatures—perhaps puppylike—that gain power over us by tapping into our emotions rather than our rational components of mind. For primates, cuteness is more than simply a disarming factor. In our robots, cuteness would ensure that humans promote their survival by taking care of them, pampering them as one might an infant, working for the benefit of their continued comfort. In short, we would develop many of the behaviors and feelings toward them that go along with social attachment. The process is reminiscent of an old *Star Trek* episode in which the away team that has beamed down to the planet below encounters a species of hamsterlike critters known only as Tribbles. Going against Spock's counsel, the team brings the harmless creatures on board the *Enterprise*, to the delight of the crew, who notice that when petted, Tribbles sing a beautiful cooing song. Before long, playing with Tribbles becomes the primary recreation of the shipmates, who are unable to resist their fuzzy appearance and soothing sounds. Meanwhile, the captain and Dr. McCoy begin to realize that petting Tribbles stimulates them to reproduce, and before long the *Enterprise* is in a desperate state, about to be overwhelmed by the exponential growth of these little fur balls. Soon Tribbles are everywhere, popping out of the food replicators, cooing from inside the ship's main computer consoles; they have reached every nook and cranny of the vessel. Only Spock, devoid of emotions, seems immune to their charms and quickly takes control of the situation by isolating these dangerously lovable creatures from the rest of the crew.

The survival of all mammals, particularly the social primates, depends critically on their ability to secure attachment and nurturance from those around them. In most primates this dependence is aimed directly at the mother, who becomes involved in a complicated species-specific exchange with her offspring, employing whatever version of "motherese" phylogeny has given her. In humans and other mammals, the exchange between parent and offspring that leads to bonding and attachment can be likened to a conversation. Even though structured language may be entirely absent in the species, a turn-taking of sorts occurs, with certain physical and behavioral characteristics of the newborn eliciting a nurturing response from the parent, which then evokes yet more stimulation from the newborn, continuing the cycle. In humans this exchange is partly composed of prelinguistic vocalizations at first, with rapid phonological development that mirrors emotional expression in the first twenty-four months.

Infants enter the world displaying a clear preference for the language spoken by their mother. For instance, studies have demonstrated that French babies as young as four days old suck a nipple more diligently when hearing French than when hearing Russian or English. Likewise, Russian newborns prefer to hear Russian rather than French or English or Italian. Detailed experiments following up on these observations showed that babies tune into the prosody (timing, stress, and inflection) of speech patterns, since playing tapes of the languages backward—which preserves most of the vowels and consonants but alters the melody—eliminates the preference. Hence, newborns are predisposed to pay attention to prosodic features of their mother's voice. Indeed, infants instinctually take such pleasure from these melodic elements of speech that they can be conditioned using prosody as a positive reinforcer (a reward) in the same way as can be done using other pleasurable experiences, such as access to its mother's milk.

These findings are less surprising when we remember that prosody conveys the emotional tone of a message. "Communication is successful," it is said, "not when hearers recognize the linguistic meaning of the utterances, but when they infer the speaker's 'meaning' from them." Many of the linguistic cues used to express intention are nonverbal. Systematic variation in pitch, tone, and duration of sounds—the music of language—is the primary venue for the infant, and it is in these signals that babies generally show the greatest interest. The newborn is naturally attracted to prosodic cues precisely because they contain the emotional meaning of speech, the very part of the message that is both critical to its social attachment with caregivers and accessible to its preverbal mind.

Interestingly, mothers across the globe—from culture to culture— speak practically identical versions of "motherese" to their infants: a complex blend of exaggerated tonal variation, eyes widened with expressive facial postures, and prodigious use of the high-toned "Hmm?" Wherever there are infants, we encounter baby talk, and it would be naive to consider this a form of linguistic instruction; the baby is certainly not enduring grammar drills. Rather, it is the innate social and emotional responsiveness of these inquisitive Lilliputian bundles that compels adults—mothers, fathers, brothers, sisters, and the rest—to speak motherese. The tiny linguophile naturally searches the faces and emotional expressions of nearby adults, an effort in identification and novelty-seeking that promotes further stimulation from parents. "The development of linguistic communication is a story about the preoccupation among the human young with things that move—faces that wrinkle, eyes that dance, voices that undulate, and hands that wiggle through the air," wrote child psychologist John Locke. "Parents obviously understand this and, correctly believing that more is better, exaggerate their facial and vocal movements when addressing their young. And to good developmental effect, for the cues to phrase boundaries are prosodic, and the cues to vocal turn taking include variations in pitch and gaze."

Why does the infant pay attention to speech? It is surely not to learn the rules of syntax, widen its semantic base, or because it thinks language is an important mode of communication. No, the process of gazing into the eyes of those around it and eliciting motherese stems rather from the child's basic, biological imperative to interact and connect emotionally with the people who nurture it. Infants orient toward the human voice, especially Mama's, and lock on to her face, studying it with deep concentration. Why should they do this? What are the biological and psychological reasons for such persistent behaviors? Surely they are adaptive in that they draw the caretaker closer to the infant, allowing it to identify those who are most likely to offer affection and nurturance.

Babies continue to learn the sounds of their mother tongue during the first year of life, all the while maintaining their innate fondness for prosody and the other features of motherese. We will see in later chapters that the infant's pleasure instinct for prosody has surprisingly long-term consequences, particularly for the evolution of aesthetic and musical preferences in the adult. For instance, synthesized sounds that have extreme pitch variations reminiscent of motherese evoke a feeling of pleasure in adults, who often associate them with happiness, interest, and surprise. Sounds that have a falling-pitch contour (high frequency decreasing to lower frequencies) elicit feelings of calm and relaxation. Imagine a parent who soothes a crying child with "Aahh," or the meditation practitioner chanting "Ohmm." Vocalizations that have a rising pitch contour have a very different effect; they tend to excite and grab our attention—"Hey!" From cross-cultural studies, it is clear that both natural and synthetic exaggerations in pitch have a universal appeal, whether they are embedded in music, speech, or song, presumably as a result of the same underlying biological mechanisms that have evolved to promote social attachment through our attraction to prosody.

An infant's face also conveys emotional information directly to the caregiver, and they are incredibly talented mimics even at birth. Developmental psychologist Andrew Meltzoff was the first to demonstrate that newborns as young as forty-five minutes old are able to reproduce facial gestures corresponding to primary emotional conditions such as disgust (tongue protrusion), surprise (mouth opening), and sadness (lip protrusion)—even before they have seen their own face! Thus from the very beginning of life, human infants are busy employing and refining their methods of communication, and the primary topic of discourse is that of emotions.

While it is true that infants enter a linguistic babbling stage, a visual analog of this behavior is seen in their tendency to produce varied facial postures shortly after birth—another sort of babbling. Through trial and error, they learn quickly which expressions evoke an emotional response in adult observers. Adults, of course, learn the same lesson, and generate a number of facial postures and behaviors, eventually stumbling on the ones that elicit facial expressions in the infant that correspond to positive emotions. Emotions, then, are the first language we use. When an adult or an infant sees an emotional expression, it instantly gains information about the displayer's current state. These talents translate to the linguistic domain, where squeaks, gurgles, and coos—the vocabulary of motherese—feed the emotional palate as well.

Studies have also shown that infants are born with a predisposition toward preferring abstract visual stimuli that look like human faces. Neonates a mere nine minutes old were shown different drawings before having ever seen a face—any face. They looked significantly longer (a common measure of preference) at a stylized pictogram of a normal human face, than at pictograms with exactly the same features but scrambled (a nose, mouth, eyes, and brows situated randomly on a circular "face"), suggesting they enter the world searching for kith and kin.

Just as prosody can be used as a pleasurable reward to condition infants, so too can the appearance of a human face. Newborns as young as two days old learn to alter their behavior (sucking and gazing) in order to maximize exposure to human faces. In fact, they master this task with astonishing efficiency, which tells us two very interesting things. First, neonates must be equipped with something that approaches single-trial learning, particularly when the task involves an evolutionarily significant variable such as the face. And second, the infant's capacity for extracting emotional and intentional information from facial features has such critical importance for survival that the pleasure instinct has made the human face a most attractive and rewarding visual stimulus for babies (and, of course, adults).

We will find in later chapters that the human face has physical properties—such as lateral symmetry and exaggerated contrasts—in common with other stimuli that infants find naturally rewarding. Our evolved pleasure instinct for these visual features has lifelong repercussions for the development of aesthetics and physical attraction in the adult. Discovering which physical features the pleasure instinct nudges us toward during our first steps as neonates will help shed light on why certain aesthetic qualities, whether they are in faces, bodies, paintings, or landscapes, are universally appealing for humans. All of these inborn talents provide the neonate with tools for establishing an emotional communion with potential caregivers. One can hardly imagine the survival benefit to an infant who routinely engages inanimate objects (either through vocal or facial expressions) with no obvious human features, to the exclusion of their brethren. Nature is unwilling to take any chances with this most critical of objectives, the biological imperative to become attached to a caregiver, receive nurturance, and eventually become enmeshed into a broader social community. In the next few chapters, we will learn how this fundamental biological rule combines with embryological and developmental processes that regulate the growth and maturation of the human brain.

Chapter 3

What Makes Sammy Dance?

*There seems to be a continuing realization by
psychologists that perhaps the white rat cannot reveal
everything there is to know about behavior.*

—Keller and Marian Breland,
The Misbehavior of Organisms

*The mind of the thoroughly well-informed man
is a dreadful thing. It is like a bric-à-brac shop, all
monsters and dust, with everything priced above its
proper value.*

—Oscar Wilde, *The Picture of Dorian Gray*

One morning in 1970 a tortured twenty-four-year-old man
with a history of drug abuse and severe depression walked into
Dr. Robert Heath's office at Tulane Medical School in New Orleans.
By then Heath was a well-known, albeit controversial, figure who
founded the Department of Psychiatry and Neurology at Tulane in
1948 after being recruited from Columbia University. Within a year
of joining the faculty, Heath and his coworkers were conducting

25

experimental studies in humans that would forever change the way psychiatrists think about emotions and at the same time provide enough source material to keep biomedical ethicists busy for decades to come.

By the time he was twenty-four years old, the patient known as B-19 had a diagnosis of temporal lobe epilepsy compounded by a history of chronic drug abuse and depression. "I live with the idea of suicide daily," he is quoted as saying, and it is reported that he made several "abortive attempts." We also learn that B-19 was homosexual and that "one aspect for the total treatment program for this patient was to explore the possibility of altering his sexual orientation through electrical stimulation of pleasure sites of the brain."

During the early years of his tenure, Heath pioneered the thera-peutic use of electrical stimulation of the brain (ESB) to treat epilepsy. Impressed by the work of Olds and Milner, who had just discovered "pleasure centers" in the brains of rats, Heath adapted their approach to recondition the brains of patients suffering from affective disorders and particularly schizophrenia. "The primary symptom of schizophre-nia isn't hallucinations or delusions," he told a reporter years later. "It's a defect in the pleasure response. Schizophrenics have a predominance of painful emotions. They function in an almost continuous state of fear or rage, fight or flight, because they don't have the pleasure to neutralize it." The idea was tantalizing—just stimulate the neural pleasure centers of a schizophrenic and this might rekindle damaged circuits affected by the disease and enable the patient to once again experience positive emotions.

Electrodes and cannulas (needle-thin tubes through which drugs may be delivered directly into the brain) were placed in fourteen subcortical structures of B-19's brain, including the septal region, hippocampus, amygdala, and hypothalamus—areas that were hypoth-esized to regulate emotions in humans and had previously been identified as locations where rats "self-stimulate."

Prior to the study, B-19's "interests, contacts, and fantasies were exclusively homosexual; heterosexual activities were repugnant to him." After B-19 recovered from the surgery, Heath and his coworkers stimulated each electrode briefly and asked their patient to report what he felt. Stimulation at most brain regions produced only mild or "neutral" feelings, and in some cases actually induced anxiety or other aversive sensations. But one electrode positioned in the septal region consistently produced an intense pleasurable response. "The patient reported feelings of pleasure, alertness, and warmth (good-will); he had feelings of sexual arousal and described a compulsion to masturbate."

During the first phase of treatment, B-19 was given a portable transistorized device that could be used to activate the different electrodes implanted in his brain. At first he experimented by stimulating a variety of sites—each time he pressed a different button, the device sent out a one-second pulse of electrical current to the corresponding electrode. Within a short time, however, the young patient was stimulating his septal electrode almost exclusively. During treatment sessions, he was permitted to wear the device for periods of three hours, and on one occasion stimulated this region more than fifteen hundred times (about once every thirteen seconds on average). During phase two of the treatment, B-19 was allowed to self-stimulate his septal electrode while watching "stag movies" of heterosexual activity, and he became "increasingly aroused." Pleased with their patient's progress, the innovative scientists hired a "lady of the evening" to assist them with phase three in which B-19 had his first "pleasurable" heterosexual encounter after being primed by five minutes of continuous septal stimulation.

ESB was used to treat hundreds of patients (not just at Tulane) through the 1970s, although with limited success in schizophrenics. An interesting observation was that patients suffering from depression or anxiety often rated septal stimulation as more pleasurable than patients without a mood disorder. At the time, it was believed that the stimulation was restoring the functioning of a weakened limbic system—a set of brain regions that regulate emotional valence in humans; however, this interpretation has been refined considerably in recent years. Neuroscientists now understand far more about the limbic system and how it communicates with neocortical structures during the expression and feeling of emotions. Most of what we know about the biology of pleasure began with an accidental discovery by two young scientists.

The Nature of "Natural" Reward

In many areas of science rapid advancement often comes from serendipitous discoveries. Working in a basement laboratory in 1954, newly doctored James Olds and graduate student Peter Milner were smoothing out the kinks of a study in which they were to implant electrodes deep into the reticular formation of rats. The German physiologist Rudolph Hess had recently shown that stimulation of the brain-stem regulates the sleep-wake cycle, and Olds believed that different sites within this region might selectively lead to either activation or inhibition of the neocortex, producing states of alertness or calm respectively.

During the first run of their experiment, each time the rat sniffed a particular corner of the square testing platform, Olds stimulated its brain, expecting that the activation would initiate the animal's natural tendency to explore and visit other corners. Strangely, just the opposite happened—the rat returned again and again to the corner where it received the stimulation. Puzzled by this, the pair soon confirmed that the electrode had not been positioned correctly in the reticular formation, as thought, but rather landed in the septal region, a largely unexplored, phylogenetically ancient part of the brain.

Realizing they were on to something important, they quickly replicated their findings and developed another experiment where each rat was allowed to directly self-stimulate its septal region by pressing a lever in a testing chamber—a twist on the classical Skinner box, where rats learn to press a lever for access to food or water. To their surprise, this produced rapid learning of the lever press response, and their rats were willing to perform a variety of tasks to have access to this stimulation. In other words, a brief electrical pulse to the septum seemed to have very similar reinforcing properties to natural rewards such as food, water, and sex. In the fifty years since this original study, self-stimulation has been found to reinforce behavior at a number of distinct but related brain regions in the limbic system and across a variety of different species, ranging from goldfish to humans. The fact that this neural circuit (and its purported function) is conserved across such diverse species suggests that it is a phylogenetically older part of the brain, having evolved in an ancestor common to all of these groups.

Perhaps no other discovery in neuroscience has created such a torrent of experiments, conferences, publications, and additional questions. As the circuit continued to be charted by the early pioneers it became apparent that brain stimulation was not only rewarding, it was also drive-inducing, and thus became a tool for studying natural motivation. Yet the big question remained unanswered: What exactly does an animal experience when its septum is stimulated? Is it pleasure? Is it sexual in nature? Or is it a general state of arousal that amplifies the natural drives of an animal depending on the contextual cues that surround it? Clearly we can't ask a rat for commentary, so we have to infer its inner state—whether we're talking about motivation, drives, feelings, or some other operational term—from its behavior. As we shall see, this is never easy.

All animals, large and small, slow and fast, have mechanisms that allow them to adapt to changes in their local environment. Even single-celled organisms use chemotaxis as a means of guiding themselves along chemical gradients toward nutrient-rich environments. We typically call these *motivated behaviors* in that they refer to any adjustments, internal or external, made by an organism in response to environmental changes. Often, these adjustments are regulatory, designed to maintain homeostasis, and they can include modifications to endocrine, autonomic, immune, or behavioral processes.

When Olds and Milner made their original discovery, motivation was largely thought to be a simple matter of drive or "need" reduction. This theoretical perspective works fine if we limit our discussion to thermoregulatory or ingestive functions—for instance, perspiring to dissipate body heat, or thirst to satisfy the need for water. However, it fails to explain other behaviors such as aggression, sex, or novelty-seeking, all of which can be triggered by an external stimulus, yet have no identifiable deficit state. It also fails to explain why in many cases normal homeostatic mechanisms can be overridden by strong external incentives, such as occurs during drug binges, while gorging ourselves on a delicious meal, or flying down a snow-capped mountain on two thin slabs of fiberglass. Instead, we typically explain these sorts of behaviors as an attraction to external stimuli or events that have appetitive or rewarding properties.

Unfortunately for most animals, food is not just splayed out for the taking; sexual partners are not lined up and waiting; and there is not always natural spring water nearby. All animals have to actively seek out these sources. Thus, motivated behavior is not simply eating a meal or engaging in sex; these consummatory responses are typically only the end points of a long and complex sequence of actions, guided in part by drives and in part by the appetitive features of an incentive. So if appetitive features of an environment—things that are inherently attractive or rewarding to animals—are not simply tied to essential needs or drives that ensure survival, how do they emerge? Are they innate or learned?

The short answer is both. Behavioral scientists have used an impressive variety of strategies for studying motivation, many of which take advantage of the fact that animals exhibit both classically conditioned (Pavlovian) reflexes and goal-directed instrumental (also called operant) behaviors. An often-used operant conditioning paradigm involves placing a hungry rat in a test chamber and making the delivery of food contingent upon some behavior. Say, for example, that whenever a small lever is pressed by the rat, a food pellet is automatically dispensed into the testing chamber. The rat, of course, has no innate or implicit knowledge of this relationship—it has to discover it by exploring the environment.

When placed in a novel situation, mammals typically exhibit a period of freezing behavior, where they stay in one place and examine their surroundings followed by a gradual increase in exploratory activity. During the exploratory phase, a rat will eventually press the lever, often accidentally, and, after a few co-occurrences of lever press–food appearance, gradually discover that a relationship exists between the two. This process is known as *associative learning*, and occurs in virtually all animals, from sea slugs to primates. It is the foundation on which most forms of learning are based. In our example, learning the link between the lever being pressed and the subsequent appearance of food is facilitated by the fact that food, in this case, is a positive reinforcer, meaning its appearance increases the likelihood of repeating the behavior that preceded it.

Understanding how associative learning works has preoccupied the minds of psychologists, philosophers, and biologists for decades. Of particular importance to the present discussion is that not all associations are learned with the same accuracy and speed. In general, associative learning occurs most easily when one of the components of the pair involves an evolutionarily important variable. For instance, rats usually learn the association between a specific food item and the sickness it induces after only a single exposure, and they use this learning to avoid these foods in the future. Such *conditioned taste aversion* has obvious benefits to the survival of an organism, allowing it to avoid ingesting dangerous and potentially lethal toxins. The same is true of fear. Rats often learn to avoid places where they have experienced foot shock after just one trial.

Contrasting this, rats typically need hundreds of trials to learn an arbitrary association between two neutral stimuli, for example an odor and a small object, if neither of the objects belongs to a broader class of stimuli that have had a significant impact on the evolution of the species. Indeed, some associations seem impossible to learn if the components are in opposition to species-specific tendencies. In their classic (and mirthful) paper "The Misbehavior of Organisms," Keller and Marion Breland reviewed a series of failed attempts at using operant conditioning techniques to teach animals to perform simple tasks. The title was a playful jab at their teacher, B. F. Skinner, whose book *The Behavior of Organisms* is widely considered a seminal work in the field of behavioral analysis.

In their first example they asked, "What makes Sammy dance?" Sammy, it turns out, is one of many adult bantam chickens that have been trained to emerge from a holding compartment and stand on a platform for twelve to fifteen seconds, after which food is automatically dispensed. In this task the only requirement for reinforcement is that each chicken must depress the platform and wait. Simple enough; however, most of the chickens developed a pronounced tendency to scratch at the platform, a behavior that became even more persistent when the waiting period was lengthened. Although the Brelands could not train chickens to perform the original task, "we were able to change our plans so as to make use of the scratch pattern, and the result was the 'dancing chicken' exhibit." The point of this article was not to demonstrate that capable trainers can outsmart poultry, but rather that after being conditioned to perform a specific response, animals can gradually drift into entirely different behaviors that seem to go directly against reinforcement contingencies. "It can clearly be seen that these particular behaviors to which the animals drift are clear-cut examples of instinctive behaviors having to do with the natural food-getting behaviors of the particular species. The dancing chicken is exhibiting the gallinaceous birds' scratch pattern that in nature often preceded ingestion." Reinforcing the chickens with food thus led to the emergence of innate behavior that anticipated, but was not in itself rewarded by, the arrival of food.

Some stimuli, as noted before, have supernormal features, meaning their reinforcing value cannot be accounted for simply in terms of drive reduction. For instance, although bland foods such as normal rat chow do not reinforce the behavior of a satiated rat, sweet foods high in natural sugars do. In fact, sugars rated most sweet-tasting by humans, such as sucrose and fructose, are stronger reinforcers of behavior in satiated rats than those ranking lower on the sweetness scale, such as lactose and glucose. Sugar is obviously an evolutionarily powerful stimulus. Knowing how to detect it in the environment and having a taste preference for it so that it would be ingested provided clear survival benefits for our hunter-gatherer ancestors who possessed these traits. We now realize the biological importance of sugar molecules as a precursor source of ATP, which powers so many of the biochemical reactions critical for cell functioning. Our ancient brethren certainly had no understanding of these processes—they needn't have for survival. For individuals to gain a selective advantage, they simply needed a hedonic preference for foods that contained digestible sugars and a means for their detection and ingestion.

Hedonic preference refers to a stimulus property that is innately reinforcing or, speaking more colloquially, naturally rewarding. Psychologists often use the terms *primary positive reinforcer* or *unconditioned reinforcer* to describe this type of stimulus, emphasizing the notion that their rewarding properties are usually in place at or before birth. Human newborns, for example, prefer sweet-tasting liquids rather than plain water immediately after birth, before ever being exposed to sugar in the external environment (a preference that facilitates breast-feeding, since mothers' milk is high in lactose). Two other primary positive reinforcers that work through taste include saltiness and a newly discovered gustatory dimension called umami, an indicator of protein content that is produced by monosodium glutamate (MSG). Both can serve as primary positive reinforcers of behavior and are discussed in the chapter on the development of taste preferences (see chapter 6).

There are primary positive reinforcers in every sensory domain— touch, smell, taste, vision, and hearing. As we shall see, many behaviors such as kin identification, parent-offspring attachment, and some forms of communication are motivated by compound reinforcers— particularly attractive combinations of primary positive reinforcers from several sensory modalities.

Another rich source of incentives that motivate complex behavior depends on *conditioned positive reinforcers* (also known as higher-order reinforcers)—initially neutral stimuli or behaviors that become rewarding through an association with a primary positive reinforcer. For instance, if our hungry rats learn that lever-pressing brings food, then this behavior itself acquires incentive value.* After learning the

---

*There is debate in the literature as to whether food or hunger actually constitutes the primary positive reinforcer in this example, since the appearance of either will increase the production of specific behaviors. For our purposes we consider food to be the primary positive reinforcer, since we are discussing it in the context of reward.

association, rats press the lever constantly, often to the exclusion of other behaviors such as exploration and grooming. However, they obviously have no innate fondness for this behavior, since before learning they seldom exhibit it, and only then at random. Clearly the incentive value of lever-pressing is contingent on the subsequent appearance of food, because this behavioral tendency terminates once food is no longer available—a phenomenon known as extinction.

Hedonic preferences, in conjunction with the thoughts, perceptions, and actions that become conditioned to them, provide fertile ground for studying pleasure and other emotions. Yet, it's important to realize that this theoretical foundation does not stem from a behaviorist school of thought, in which we are born as blank slates waiting to be writ upon. Quite the contrary, it assumes there are evolutionary and developmental constraints that shape *what we are likely to learn, when we are likely to learn it, and how such learning takes place.* Before we discover the ways in which each sensory modality contributes to the hedonic palate, however, we will first examine how this ancient circuitry evolved in our species and consider how the process parallels the embryological development of the human nervous system.

Essential Hardware

Since soft tissue does not fossilize, our understanding of human brain evolution comes mainly from comparative studies of living species that are closely related to *Homo sapiens.* There have been discoveries of early primate skulls that happened to fossilize with an imprint of the former owner's brain, allowing an endocast to be made of the cortical surface; however, artifacts such as these tell us nothing about the inner circuitry of these ancient brains and very little about their gross anatomy. Instead, scientists have used a different approach.

The relationships among the major vertebrate classes have long ago been organized by examining the anatomical characteristics that distinguish species. This system of classification has been refined through the years by making further comparisons based on fossilized bone remains and DNA fragments in an effort to reconstruct the phylogenetic history of living forms. Powerful new methods have been used recently to compare DNA fragments of living animals from different branches of the phylogenetic tree and have found that in many cases (but not all) the phylogenetic reconstructions based on genetic evidence correspond well with those based on studies of fossilized remains.

These reconstructions can be used to determine whether similarities between two animals are the result of a shared evolutionary history or rather co-evolved independently in both species. For example, since chimpanzees, apes, and humans evolved from a common ancestor some four million years ago, it is likely that the brain structures common to all of these species were present in that ancestor. Brain regions that are not common to all three species are most likely to be evolutionarily newer structures derived from earlier predecessors. Similarly, brain areas that are common to all Old World monkeys (for example, Macaca) and hominoids (for example, apes and humans) are most likely even older in that they stem from the common ancestor class of Old World anthropoids that existed more than twelve million years ago.

Understanding approximately when (in terms of evolution) a brain region came into existence helps us in several ways. First, if we can show that other anatomical changes co-appeared at the same time as the brain region in question—for instance, the development of bipedalism or forward-facing eyes—it may be possible to reconstruct the selection factors that contributed to its evolution. Second, since we have a growing understanding of the relationship between brain and behavior, information about the general evolutionary history of the human brain allows us to develop theories about the evolutionary history of our cognitive and emotional capacities.

The evolution of mammals from reptilelike ancestors about three hundred million years ago brought the gradual accumulation of several distinctive features: the appearance of hair; sweat glands; mammary glands and suckling behavior; specialized teeth for grinding, slicing, and piercing new food sources; and physiological mechanisms for maintaining a constant body temperature (thermoregulation). All of these adaptations suggest that early mammals led a predatory lifestyle. During this period, the brain also went through a number of significant transformations.

Chief among these is the development of the neocortex, the vast outer portion of the cerebrum that constitutes about 85 percent of the human brain's total mass and that is responsible for higher-level cognitive functions such as language, learning, memory, and complex thought. Reptiles and birds possess an anatomically simpler, three-layered version that is sometimes referred to as archicortex (old cortex). Mammals have a six-layered version that is far denser in cell counts per volume and contains a greater diversity of cell types; however, they also possess several three-layered cortical structures, such as the olfactory cortex and hippocampus (collectively referred to as allocortex), which are similar to the archicortex of nonmammals. For these reasons, it is believed by many (but certainly not all) neuroanatomists that the neocortex is unique to mammals and derived from phylogenetically older cortical areas.

This evolutionary ordering of three-layered cortical areas predating the six-layered neocortex is paralleled in the embryological development of humans. By the sixth week of gestation, a human fetus will already have many brain-stem structures partially developed. These will control basic physiological functioning in the newborn— respiration, sleep-wake cycles, thermoregulation, and a host of motivated behaviors. Brain-stem sites are followed by the initial appearance of subcortical regions such as the thalamus and hypothalamus sometime around the tenth week. As we shall see, many of these brain regions play a pivotal role in the generation of motivated behaviors and act as sensory integration sites. By fourteen weeks the allocortical areas start to develop, eventually becoming part of the limbic system, a region responsible for learning, memory, and processing emotional information. By the sixteenth week, cells begin to appear in what will eventually become the neocortex. However, it's important to note that none of these areas is wired together functionally yet. This occurs in two distinct stages—neurogenesis and synaptogenesis.

How the Developing Brain Gets Wired

Our understanding of how the human brain develops from a smooth sheet of ectoderm* into the mature adult form has changed radically in just the past ten years. Scientists now have a much deeper appreciation for the way developing brains depend on specific types of stimulation patterns and sensory experiences to activate important genes. These insights have come from studies of developing infants using noninvasive brain imaging techniques, neuropsychological experiments, and an improved understanding of how genes actually work.

Genes have two basic components, a template (or coding) region that provides information about how to make a specific protein, and a regulatory region that determines when a gene is expressed or repressed. The template region is what we usually think of when we hear the phrase "it's in our genes." The information in this region is not modified by experience or learning, only through mutations, which are rare and essentially random. The regulatory region, on the other hand—the on/off switch of the gene—is highly sensitive to a host of experiential factors.

A great variety of signaling proteins can bind to the regulatory region of a gene and modulate its subsequent transcription and expression. These signaling proteins are known as transcription regulators. Put simply, when transcription regulators bind to a segment of a gene, they activate its expression and the eventual production of a new protein. Thus they have direct control over whether a gene is turned on or off.

A number of factors influence the way a transcription regulator binds to a gene. Both internal and external stimuli (that is, things we experience) activate signaling pathways that result in alterations of this binding process. Some signaling pathways are differentially activated as a result of the normal developmental process. Others are activated by stress, learning, hormonal changes, or social/experiential interactions.

---

*Ectoderm is the outermost of the three primary germ layers of an embryo, from which the epidermis, nervous tissue, and, in vertebrates, sense organs develop.

For example, psychological and physical stress causes the release of the adrenal gland steroid glucocorticoid (also known as cortisol), which circulates in the peripheral and central nervous system (brain and spinal cord). In the brain, this steroid activates a transcription regulator that binds to the regulatory region of several genes, inducing the transcription and expression of new proteins involved in the long-term regulation of the stress response. Hence, social factors such as stress regulate gene expression and the subsequent production of specific proteins. Indeed, all forms of learning are incorporated into our biological makeup in the altered expression of specific genes that encode the production of selective proteins in brain cells.

Gene expression can be extremely selective in targeting the production of proteins unique to a specific type of nerve cell and brain region. A particular experience, say a psychological stressor, will result in the production of a very specific set of proteins, while another experience, for example, learning a new phone number, will result in a different set. These experience-induced changes in gene expression and subsequent protein production are not transmitted from generation to generation genetically. None of these alterations in gene expression is incorporated into the sperm or egg, and therefore are not heritable. All changes in gene expression that result from learning or being exposed to experiential/environmental factors are transmitted culturally rather than genetically, and they clearly have a profound impact on the way brains develop.

The fact that genes have essentially two functional components has important implications for development and the relationship between nature and nurture. Being familiar with the way genes really work makes it easier to see why most biologists have long ago given up on the nature versus nurture debate as a false dichotomy. The genes that code the way brains are built do not contribute to development unless they are transcribed and expressed. Hence, experience is an essential part of development even at the level of the gene.

How do genes build brains? As I write this chapter my wife is four and a half months pregnant, and every day brings new questions about little Kai's development. The typical adult human brain has about 100 billion nerve cells or neurons. Each neuron connects to thousands of others, resulting in about 10^{14} (a 1 followed by 15 zeroes) different connections. How, then, do the 25,000 or so genes identified by the Human Genome Project code such a combinatorically large and complicated system? Clearly, since the numerical differences are so great, genetic information does not uniquely specify where every single neuron resides or where each of its thousands of connections will terminate. Instead, genes specify more general rules for neuron development and migration.

At the earliest stages of development, we start off as three primitive cell layers: endoderm, which consists of cells that eventually line our internal organs and vessels; mesoderm, destined to become the major structural components of the body, including bones and muscle groups; and finally, ectoderm, which becomes the central nervous system, skin, hair, and nails. Kai's entire mental world—his thoughts, emotions, sensations, and perceptions—emerge from this thin sheet of cells. Within the first few days of gestation, the primitive cell layers elongate and fold into a cylindrical tube called the notochord, the progenitor of Kai's spinal column. This process recapitulates the earliest event in the evolutionary transition from invertebrate to vertebrate forms that occurred more than 600 million years ago.

Once the notochord is formed, it guides the ectoderm layer, which progresses through a series of well-defined stages, first thickening and then folding in on itself to form the neural tube. At about nineteen days into gestation, just about the time when Melissa and I first learn she is pregnant, the earliest form of Kai's future brain and spinal cord begin to emerge through a process called neurogenesis. During this period, the front end of Kai's neural tube develops three enlargements, which eventually become the two cerebral hemispheres and the brain-stem. The neural tube then goes through a rapid growth spurt, where the entire cycle from cell division to cell division takes place in about an hour and a half. Some of these precursor cells are destined to become neurons, while others will mature into glial cells, which serve a variety of supportive functions in the brain.

As cell division and replication continue, the three enlargements begin to take on more detail, eventually forming all the major components of Kai's brain. At two months into gestation he is little more than two inches long, yet all of his major brain structures have begun to take shape, including the elementary forms of the medulla, pons, and midbrain, which combine to form the brain-stem; subcortical structures such as the thalamus, hypothalamus, and basal ganglia; then a bit more slowly, the allocortex; and even more slowly, the neocortical regions. Kai's brain will develop from the bottom to the top, with lower brain-stem structures such as the medulla maturing first, followed in sequence by the upper brain-stem, subcortical areas, the allocortical regions, and then the neocortex.

It's often asked why neurogenesis starts at the bottom of the brain and progresses toward the top (or since the brain is three-dimensional, from the inside of the brain to outer regions). A clue can be found if we compare the embryological development of an individual with the evolutionary development of our species. Until roughly four weeks of gestation, the embryo that will become Kai is practically indistinguishable from embryos of many bird, reptilian, and mammalian species. But by the sixth week, he begins to look more and more like a mammal, and by week seven he appears decidedly primate. As a general rule, species of similar phylogenetic forms tend to look alike for longer periods during development. For example, human and chimpanzee embryos share strikingly similar features until about the seventh week of gestation, at which time they begin to diverge in appearance. Human and rabbit embryos, on the other hand, share similar developmental features only until about the fourth week of gestation. Thus, embryological development is highly conserved across different species—those structures appearing earliest in development are most common across species and thought to be derived from phylogenetically earlier forms.

This makes sense when one considers the vast diversity of adult life forms that mature from a single fertilized egg. Evolution is driven by the natural selection of forms that are altered through mutations, and chances are much greater that an offspring will be viable if a mutation occurs at a developmentally later rather than earlier stage. Mutations that occur early in embryological growth tend to have devastating consequences because they alter all subsequent stages of development. Consequently, successful adaptations are most often "added on" or modified from existing structures that were present in earlier phylogenetic forms.

This relationship between ontogeny and phylogeny explains why human brains take so long to develop and why certain regions mature before others. The human brain, particularly certain areas of the neocortex, distinguishes us from other primates and takes significantly longer to develop and mature than other systems, such as those that control respiration and circulation. First to come online are those brain regions that are involved in the essential behaviors common to most mammals—for example, simple reflexive movements and those that eventually control respiration and digestion. Many of these simpler behaviors depend on neurons in the spinal cord or lower brain-stem.

By the end of the first trimester, Kai has a fairly mature brain-stem and diencephalon, which consists of the thalamus and hypothalamus. The thalamus is a critical sensory relay station that takes part in a two-way dialogue with different areas of the neocortex, sending new information through and often being instructed by the neocortex in a top-down manner to filter certain features of sensory information so that it never reaches conscious perception. The hypothalamus is just as important because it will serve as the mediator between the rest of Kai's brain and his endocrine and immune systems. His first sensations of hunger and satiety, and the emotions that accompany these survival behaviors, will be mediated largely by cells in his newly formed hypothalamus.

Even within a specific sensory modality—sight, for example— those features of vision that are most common across species, such as detecting movement and contrasting light intensity, arise functionally well before other features, such as color vision and depth perception, which are more unique to primates. As mentioned above, this is because the brain regions that control each of these sensory functions arise in a specific sequence laid down by our genetic programming that reflects a gradient from functions common to many species to functions more unique across phylogenetic classes. We start off broad functionally, and with development continue to add those features that are first common to all mammals, then those that are only seen in primates, and finally those that only humans possess.

After he is born, Kai will be able to see things that move long before he'll be able to see colors. This is because he will emerge from the womb with only brain-stem, thalamus, and the most primitive visual cortical regions connected and working. And as will become clear, these circuits are in a very sensitive state. For the processes of neurogenesis and synaptogenesis to continue the job of connecting his entire visual brain and fine-tuning it, the circuits he has thus far must be stimulated by specific types of visual experience. It is in the service of this developmental necessity that pleasure—driven by the hypothalamus and other limbic structures that come online early—participates to ensure that newborns seek out the forms of sensory experience that optimally stimulate the time-sensitive growth and maturation of their brains.

This can be thought of as a "bootstrapping" mechanism, whereby developmentally early brain systems that support essential survival and regulatory processes contain the functions that ensure the further development of higher-order brain systems involved in more complicated forms of perception and cognition. Just as Kai's brain will need nutritionally rich sources of metabolic energy to continue normal development, so too will it need to encounter sensory-rich stimulation at specific intervals during maturation. The experience of pleasure thus provides newborns with a general rule for seeking out patterns of stimulation that guide the normal development of all higher cognitive functions, including perception, language, and abstract reasoning—functions that have traditionally been treated as being entirely separate from their more passionate brethren.

At four and a half months into our adventure, Melissa—a prepregnancy vegetarian—can now keep her food down, and occasionally sends me out (often very late!) to pick up a Whopper with extra pickles. It is at about this time that the processes of neurogenesis and cell proliferation begin to peak in Kai's brain. The numbers are staggering. To reach the estimated one hundred billion neurons that comprise a human brain—the majority of which are in place by midgestation—Kai will have to produce an average of five hundred thousand cells per minute during the first four and a half months. During this time of cell proliferation, neurons migrate to their intended locations, and once there begin to grow extensions toward other cells in a process called synaptogenesis.

One hundred billion neurons alone do not make a brain. It's in the detailed wiring that connects cells together within and between different regions that we find the mechanisms of sight, smell, hearing, touch, and the multitude of unique capacities that make us human. Each neuron has three basic parts—the cell body, which contains most of the metabolic machinery; dendrites, which take in information from other cells; and an axon, which carries information from the cell to target cells with which it will communicate. After neurogenesis, brain cells extend their dendrites and axons, and form synaptic connections with other cells. A synapse is essentially a point of communication between two neurons, and the dialogue is electrochemical. So important is this process to life that its absence, as measured by a lack of significant electrical activity in scalp EEG recordings, is taken as a definition of death.

It's been estimated that between midgestation and two years of postnatal life—the peak period of synaptogenesis—close to 15,000 synapses are produced on every cortical neuron. This averages to an astounding 1.8 million new synapses formed every second during this period. At present, there are several controversial theories as to how a cell finds its correct target cells. For example, how do the output cells of the retina know they must bypass certain brain regions and terminate their axons in the proper portions of the visual thalamus? And how then do those thalamic cells know to project to specific regions of the primary visual cortex? Although no single theory is consistent with all of the available data, it is widely accepted among neurobiologists that synaptogenesis is a comparatively long process—continuing through gestation and several years of postnatal life—and works primarily through competition.

Nature begins the job. The general wiring scheme is laid down by our genetic programming and follows a very specific developmental sequence of time-dependent growth patterns that resembles that observed during neurogenesis. The first areas to undergo significant synaptogenesis are in the lower brain-stem, followed by the upper brain-stem, the diencephalon, various subcortical regions, the allo-cortex, and finally neocortical regions. Neuroscientists are only beginning to understand how genes code the basic wiring of the brain, and the ways in which this process depends on experience. Genes, it turns out, direct axons and dendrites only to their approximate target locations. Once these beginner circuits start to function, however, experience and the pleasure instinct take over to fine-tune the connections and shape them into a precise network unique to a child's environment.

From the end of gestation through early childhood, Kai will produce about twice as many synaptic connections as he will eventually need in his adult brain. Cells make and receive thousands of synapses during this promiscuous period, resulting in a large-scale but rather diffuse communication system between different brain regions. The over-production of synapses is an evolutionary trade-off—since the small number of genes we possess simply cannot dictate each of the qua-drillion or so connections between brain cells, they instead provide general rules for making contacts between areas. Once these diffuse connections are in place, nature takes over and begins to selectively prune certain synapses based on the types of stimulation patterns (that is, experiences) the brain receives. In this way, experience, guided by the pleasure instinct, refines the communication network within the brain, making certain connections stronger while removing others.

Synaptic pruning is a requirement of normal brain development. As I mentioned, it works directly through a process of competition—survival of the fittest. Each of the synaptic connections in Kai's brain has the potential to survive past this competitive period of pruning, but about half will perish, and along with them certain functional capabilities. The rule is simple—use it or lose it. To survive, the syn-apse between two cells must be activated consistently. Those synapses that are activated the most have three advantages over less active synapses. First, when activated, they tend to inhibit surrounding synapses (through cellular responses) that are competing for stimulation. Second, the electrochemical dialogue between active cells triggers a series of biochemical reactions that strengthen the synapse, cementing it in place. And third, synapses that are inhibited during this competitive period have an increased likelihood of triggering active processes that weaken and eventually terminate the connection. These fundamental discoveries from neuroscience have profound implications for the way we raise our children and educate our young during the first two decades of life.

While the major work of neurogenesis and synaptogenesis is finished relatively early during development, synaptic pruning occurs very slowly, lasting at least until an individual's early twenties and perhaps longer. Although it takes decades, synaptic pruning generally follows the same sequence as neurogenesis and synaptogenesis, progressing from lower structures of the brain-stem that tend to be more broadly represented across phylogenetic class, upward toward structures such as the prefrontal cortex that are most unique to primates. One consequence of this sequential development is that the behavioral, perceptual, and cognitive functions that depend on each of these regions also tend to arise through a specific—and culturally independent—sequence.

Once a particular brain region undergoes synaptogenesis and the overproduction of synaptic connections, this marks the onset of abilities regulated by that region, such as the emergence of color vision, sound localization, and language. Although experience influences all stages of development, it is predominantly the long period of synaptic pruning that fixes the overall quality and nuances of these abilities. As synaptic pruning refines certain abilities, for example, a capacity to discern phrase boundaries unique to human languages, the process also results in the sacrifice of alternative synaptic connections that gradually make it difficult to acquire other abilities, such as the perception of sounds that are not normally a part of human experience. Hence, synaptic pruning is essentially a selection process driven by experience.

As we shall see, the behaviors that emerge during synaptic pruning—guided by the pleasure instinct—are really forms of self-stimulation that infants, toddlers, children, and adolescents *must* produce to ensure normal brain development given the ecological and environmental context in which that development and growth take place. A central theme of this book is that nature has solved this functional requirement in primates (and perhaps other mammals) by exploiting the early capacities of limbic structures to produce pleasurable sensations, which have been associated through our long evolutionary lineage with optimal forms of brain stimulation.

From culture to culture, it is striking that infants and toddlers pass through the same developmental milestones, many of which can be used to gauge how far their brain has matured. Once a potential capacity arises, they must self-stimulate in ways that encourage the further development of the brain systems that mediate these and related behaviors. We find this process at work in every sensory modality, and the sequence of self-stimulating behaviors that emerge reflects the order in which our primary sensory systems come online during development. As we will learn in the chapters that follow, this developmental progression is an echo of our evolutionary past.

Part Two

The Pleasures of the Sensory World

Chapter 4

The Pleasure of Touch

*Touch is the parent of our eyes, ears, nose,
and mouth.*

—Ashley Montagu, *Touching*

Touch seems to be as essential as sunlight.
—Diane Ackerman, *A Natural History of the Senses*

The sign above the door declared that the children inside were
"unsalvageable." By the time the world met him, Izidor had lived at
the Hospital for Irrecoverable Children in the northwest mountain
town of Sighetu Marmatiei, Romania, for most of his eleven years.
Crippled by polio, he was abandoned shortly after birth and ware-
housed with five hundred other children, all victims of Romanian
president Nicolae Ceausescu's campaign to expand the population
of the country. Ceausescu saw the economic problems of his land as
ones that stemmed from a shortage of labor. His solution was to institute a strict program of population control, and in 1966 he banned
abortion, birth control, and divorce, and decreed that all Romanian
women must bear five children apiece. As a result of this program,

state-run orphanages were soon overflowing, and by the late 1980s, Romania had about 180,000 abandoned children in a country with a population of only 5 million.

In 1990, the ABC News program *20/20* did a story called "The Shame of a Nation," about the plight of Romanian orphans. Horrific images of anguished children in tattered clothing, sleeping on urine-soaked mattresses and floors, mobilized an international relief effort that continues today. In the years that followed, parents in Western Europe and America who adopted some of the orphans began to notice an odd syndrome of autistic-like developmental problems in these otherwise healthy children that included self-hugging behavior, rocking, and extreme sensitivity to touch. Behaviors such as these have been observed in institutionalized children for hundreds of years, but it was not until the late nineteenth century that pediatricians began to study seriously the effects of institutionalization on child development.

During the 1800s, more than half of all American infants institutionalized in foundling homes died of a mysterious disease called marasmus (a Greek word meaning "wasting away") before their first birthday. As late as 1915, the infant mortality rate approached 100 percent in many homes, and during this year the distinguished pediatrician Henry Dwight Chapin presented alarming findings to the American Pediatric Society. His data showed that nearly every foundling home sampled from ten major cities in the United States reported an astonishing 100 percent mortality rate for children before their second birthday. Once Chapin's findings were made public, reform measures were implemented that, for the most part, called for improved access to breast milk through wet nurses. This strategy decreased the mortality rate somewhat at most institutions, but Chapin and others were stunned to realize that even those foundling homes with first-class medical care and nutrition had mortality rates that were comparable to, and in some cases actually worse than, less-equipped homes. The question was why.

It is important to note that at the time, American physicians and fashionable parents were under the spell of Luther Emmett Holt, the influential professor of pediatrics at Columbia University who recommended a system of child rearing that most parents would probably not feel comfortable with today. He warned against cradling and rocking infants and advised that parents should handle their babies—including newborns—sparingly. Chapin began to suspect that it was just this type of emotional aridity that was contributing to marasmus and high infant mortality rates.

However, it wasn't until the late 1930s that the causes of marasmus were identified. A clue came from the studies of Chapin and other pediatricians who noted that the disease was just as prevalent or perhaps even more so in the "best" homes and institutions, where medical care and hygiene were superior to those homes with smaller budgets. A few insightful pediatricians noticed that those institutions that occasionally fostered children out to single families on a temporary basis because of budgetary constraints often had fewer cases of the disease. The obvious conclusion was that it was the increased physical contact in the form of gentle handling, rocking, massage, cuddling, and bathing that seemed to benefit the health of the temporarily fostered infants. Before long, the practice of mothering was incorporated into the daily regimen of most institutionalized infants. The pediatric wards of the famed Bellevue Hospital in New York reported a nearly 25 percent decrease in infant mortality in a single year after mothering sessions—in which infants were gently stroked, massaged, and rocked—were formally introduced into their child care curriculum.

A striking similarity exists in the descriptions of nineteenth-century children reared in foundling homes and those institutionalized in modern Romania. The repetitive self-hugging and rocking movements, the extreme sensitivity to tactile stimulation, and the flat affect of many of these children are all characteristics that would have seemed familiar to Izidor. In his autobiography *Abandoned for Life*, he describes the morning-to-evening routine at the Sighetu Marmatiei orphanage: "Most of the time, the children did the same old thing; rock back and forth, sleep or hurt themselves. If they continued hurting themselves, the house nannies either gave the children more medication or put them into straight jackets." With a ratio of one nanny for every 75 children during the day and one for every 150 children during the evening, there was not much time for the kinds of tactile stimulation—simple, loving touch—that are known to be so crucial for normal psychological and physical development. It is only in the past twenty years or so that scientists have begun to show that pleasurable touch during the early period of life is crucial for normal brain development and why its deprivation can produce autistic-like behaviors.

Touch is often called the "mother of the senses" because it is believed to have been the first sensory ability to have evolved—the sense upon which all others have been based. Well before the appearance of primates and even before mammals, touch existed in the earliest unicellular organisms, such as the predecessors of the modern coliform bacteria *E. coli*. This phylogenetic observation has not been lost on developmental biologists, who point out that the skin is the first organ to develop embryologically in complex, multicellular organisms, and it is their first portal of communication. Touch, the sensation that is most intimately associated with the skin, is the earliest sense to develop in all species of mammals, birds, and fish. Before a primate embryo is even an inch long from head to rump—long before it has eyes and ears—it will respond to gentle stroking of its lips and outer nose by extending its trunk away from the source of stimulation.

In the human embryo, tactile reactivity begins at about six weeks of gestation. As with all primates, a human embryo first responds to touch around its lips and nose, and gradually, with further development, begins to sense touch on other parts of its body. By nine weeks of gestation it will respond to touch of its fingers and hands, and by twelve weeks it will curl its toes if the soles of its feet are stroked. This well-known head-to-toe progression occurs in all primates and results from two important forces involving brain development.

First, lower brain-stem regions develop and begin to function well before neocortical areas of the brain come online. This growth process limits the sensory and behavioral capabilities of the newborn to those supported by the brain regions that are developed and connected enough to show some degree of functionality. The reflexive turn of the embryo at six weeks of gestation is controlled predominantly by brain-stem circuits without any of the information being passed on to "higher" neocortical sites. In other words, the embryo will react when its mouth or nose is stroked at six weeks of gestation, but it will have no conscious perception of the experience. For touch to be perceived, the embryo will have to wait an additional six weeks for those higher sites to develop and begin to exhibit even the most basic physiological functions.

The second constraint that influences the head-to-toe development of touch sensation is found in the way bodily sensations are mapped in the brain and how this organization changes with experience. Anyone who has taken an introductory psychology course has seen the grotesquely shaped homunculus with its exaggerated lips, nose, hands, and genitalia. The relative size of each body part is drawn to be proportional in size to the amount of cortex devoted to processing information from that area. The point is, not all portions of the body have equal representation in the brain—it is not a democracy. To understand how brain maps of the sensory world develop and how this process influences behavior, we should first think about how touch sensation works.

Sensations

Physical sensations arise from the stimulation of a variety of receptors distributed throughout the body. Although we often think of touch as tactile sensation, it actually has four forms: discriminative touch, which is used to identify the shape, size, and texture of objects and their movement across the skin; proprioception, which gives us a sense of body position and movement through space; nociception, which signals that tissue has been damaged or irritated and is perceived as pain or itch; and temperature, which indicates relative warmth and coolness. Each of these sensory modalities is regulated by a distinct set of receptors and brain circuits that comprise the somatosensory or touch system.

When we talk about touch, we are typically referring to the most colloquial use of the term, which implies tactile sensations. By the time my son is old enough to grab his first teething ring, his sense of touch will be far more developed than any of his other senses. As he squeezes the ring, touch and temperature receptors embedded in his skin become activated and originate electrochemical signals that journey up his spinal cord to his brain-stem. From there, the signals travel to his thalamus and finally his somatosensory cortex, where the activation results in the conscious perception of the tactile and temperature qualities of the ring.

But all of this is still months away because Kai is a fetus, and his somatosensory cortex is still in its earliest stages of development. Eventually this region of his brain will become a highly ordered map of his skin surface. But this process will depend on his ability to stimulate these early circuits. A very active area of research over the past twenty years has been aimed at understanding how early life experiences shape the way these topographical maps form in the brain.

We've known for some time that brain map development is highly sensitive to experience, particularly during the periods of synaptogenesis and synaptic pruning. Psychologist William Greenough has popularized the idea that brain maps are shaped by two predominating forces during development (and probably even in adulthood)—those that are experience-expectant and those that are experience-dependent. Experience-expectant interactions are those forms of stimulation that all humans must experience to ensure normal brain development. Experience-expectant stimuli fine-tune the brain during major growth periods, and in a very real sense pick up where the primate genome, limited in its capacity to precisely specify each developmental detail, leaves off. In contrast, experience-dependent interactions involve experiences that are unique to individuals—for example, information about their personal identity, familial structure, and the particular social mores that exist in their community.

During synaptogenesis and the prolonged period of synaptic pruning, the brain is an experience-expectant organ par excellence.

Mammalian genetic code provides only enough information to build a beginner mapping between body surface and each somatosensory brain region; the rest of the job depends on stimulation. Mice, for instance, use their whiskers to sense objects and each other much like primates use their hands. A mouse's cerebral cortex contains a very detailed somatosensory map of its whisker region—topographically organized into rows that correspond to the way its whiskers are organized on its face. Each whisker has a small cortical region shaped like a barrel (hence they are referred to as cortical barrels) that forms in the first few days after birth. If, however, a whisker follicle is removed during this period, the corresponding cortical barrel fails to develop. Instead, adjacent cortical barrels that correspond to adjacent whiskers encroach and take over the space that was once devoted to the plucked whisker. Thus, whisker sensation is required for the normal development of whisker representation in the mouse brain.

Interestingly, if a select group of whiskers is given additional tactile stimulation during the first five days of life, their associated cortical barrels grow larger than those associated with less used whiskers. Hence, extra stimulation of specific whiskers leads to a larger portion of the brain devoted to processing their tactile information. Together, these experiments demonstrate the critical importance of sensory experience during brain development. Lack of stimulation during a critical period can lead to stunted growth of the corresponding brain region and a loss of function, while extra stimulation can facilitate growth.

An interesting finding with practical implications from the past few years is that stimulation in one sensory form, such as touch, often leads to improvements in other areas, such as learning and memory, and these effects can last into adulthood. For instance, when mice are placed in enriched environments filled with interesting objects to scramble over and make contact with, their cortical whisker barrels grow significantly larger than those of mice from control groups living in sparser conditions without increased opportunities for stimulation. Surprisingly, enriched mice also end up being "smarter" adults compared to their impoverished cousins even if they were exposed to the enriched environment for only a brief time. Adult mice that were raised in enriched environments for durations as short as one month as juveniles perform far better than mice reared in more sterile conditions on practically every task designed to assess learning ability in rodents.

Additional studies have since shown a second important observation. When young mice are given a choice between an enriched environment (with toys, bedding, water, and chow) versus a typical environment consisting simply of bedding, water, and chow, the vast majority spend significantly longer periods of time in the enriched environment. This effect persists into adulthood, but becomes less significant in aged animals. These studies tell us that rodents have an innate preference for enriched surroundings where more opportunities for stimulation exist, and that this type of stimulation improves brain growth and functioning.

Of course, any parent will tell you the same is true of us primates. Hedonic preferences have evolved in every sensory modality to nudge us toward environments and behaviors that satisfy the experience-expectant requirement for normal brain development. Ultimately, the sequence of behaviors that emerge in human infants, toddlers, and throughout childhood is a guide to what experiences the brain needs to fine-tune itself and function effectively in the particular ecological niche in which its owner resides. To this end, one might ask why certain forms of touch sensation are pleasurable and clearly preferred over others, and how these kinds of experiences help wire the brain.

The Evolution of the Rocking Chair

The kinds of experiences sought out by newborns, toddlers, and children are not simply a random collection of idiosyncratic tendencies that vary from individual to individual. Rather, the preferences for certain forms of stimulation (such as motion), and the periods during development when they emerge, are fairly consistent across individuals and cultures. This is exactly what one would expect if these patterns of stimulation are critical for the development of brain systems involved in touch, and these systems, in turn, have provided selective advantages during the evolution of our species.

While early touch sensation is required for normal somatosensory development, it is unclear how long the critical period lasts in humans. Some recent studies have shown that the brain systems involved in touch continue to change as a result of experience well into adulthood. This makes intuitive sense, since adult primates can obviously learn new information and improve their sensory acuity with training. The implication of this is that adult brains are still plastic (although perhaps not as malleable as those residing in younger bodies), and that the need for specific kinds of sensory experiences continues to be important for brain maturation well beyond the early formative years.

An example of this process can be found in our love of one particularly pleasurable somatosensory experience—the feeling of being in motion. Proprioception, the sense of the position and movement of one's body, depends on signals from the skin surface as well as from muscles and joints. These signals travel through separate circuits in the lowest portions of the brain-stem and thalamus but eventually converge in the cerebral cortex with signals from the vestibular system. The vestibular system monitors changes in head and body posture relative to the Earth's gravitational pull and an organism's direction of motion. Since all organisms have had to orient themselves relative to these two elements during evolution, the vestibular system is thought to be as phylogenetically old as other components of the somatosensory system.

Vestibular functioning begins at about the same embryonic period in primates as touch sensation. Even though Kai is still three months away from being born, his proprioceptive and vestibular systems are mature enough to function and send signals that converge in his prenatal cerebral cortex. This provides his earliest sensations of motion. It is striking that the integration of these systems begins to develop at about the same time that Kai's general activity level cranks up to new extremes. Each night as Melissa settles down for the evening, he begins his prenatal gymnastics—head, leg, and arm flexion and extensions produced with seemingly endless repetition. By this point in gestation, Kai also attempts to compensate for sudden changes in his mother's position—when she stands up or rolls over—by rapidly extending his arms and legs, a response known as Moro's reflex. After he is born, our son—like all newborns—will quickly succumb to the pleasures of motion. His earliest sensory satisfaction will be found in the touch and warmth he feels when being gently caressed, and the soothing comfort he'll find in motion will run a close second.

Babies enjoy the sensation of motion from the moment they are born. Newborns are pacified by rocking, gentle swaying movements, and by being carried around the house, while toddlers enjoy virtually all forms of repetitive motion—particularly jumping up and down for hours in those baby bouncers. Older children graduate to more sophisticated methods of satisfying their thirst for motion—carousels, tricycles followed by bicycles, skating; the list goes on and on. Humans of all ages seem to have an innate fondness for motion, but the way this desire is manifested clearly depends on age and other developmental factors, as well as on cultural norms. The inborn pleasure we take in motion can clearly have detrimental effects when satisfied in the "wrong" way, such as by speeding or other forms of thrill-seeking that endanger personal safety.

We're often told that newborns and infants are soothed by rocking because this motion emulates what they experienced in the womb, and that they must take comfort in this familiar feeling. This may be true; however, to date there are no compelling data that demonstrate a significant relationship between the amount of time a mother moves during gestation and her newborn's response to rocking. Just as plausible is the idea that newborns come to associate gentle rocking with being fed. Parents understand that rocking quiets a newborn, and they very often provide gentle, repetitive movement during feeding.

Since the appearance of food is a primary reinforcer, newborns may acquire a fondness for motion because they have been conditioned through a process of associative learning (see chapter 3).

Another possibility is that the sensation of motion is actually a primary reinforcer itself and is inherently pleasurable independent of any associations it may have with other forms of stimulation. In this scenario, the sensation of motion is desirable and soothing because it is critical for primates to experience these types of stimulation for normal development. The fact that both motion and touch sensation are the earliest sensory capacities to emerge and are keenly dependent on stimulation during development suggest that newborns, toddlers, and children may seek out experiences of motion as a primary reinforcer that promotes the continued growth and maturation of these systems. Children, of course, are unaware of this relationship—they needn't be consciously aware of why they enjoy certain forms of movement for this growth and maturation to take place. They only need to crave the experience enough so that they seek it out. This process is reminiscent of the relationship between our fondness for sweets and their ability to produce energy. Evolutionary mechanisms do not require that primates understand the biochemical reactions involved in converting sugars to ATP for them to benefit from the relationship. They only have to seek out sugar (for whatever reason) to give those individuals in a population who crave these high-energy food sources a selective advantage over their competing hominids. The end result is a contemporary population of sweet-toothed primates with an extraordinary means for fulfilling their biologically driven desires.

How Movement and Touch Promote Brain Development

The early emergence of touch and motion sensation is critical to the normal development of other parts of the nervous system. Because these sensory capacities have such an early onset, they organize other sensory and motor systems. Biologists have been moved by these findings to begin thinking of development as a cumulative process rather than a simple schedule of programmed events. For instance, children who have delayed vestibular system maturation tend to reach motor milestones such as crawling, sitting up, and walking more slowly than normal children, sometimes taking twice as long to pass these hurdles. Each of these behaviors depends on a sense of balance, and hence vestibular function becomes a critical force in organizing motor behavior.

Children born with deficits in touch and vestibular function also frequently have emotional and cognitive disturbances that often involve learning and memory, attention, visual-motor integration, language, and autism, to name just a partial list. Neuroscientists sensitive to this new view of development are now beginning to understand how early touch and vestibular experiences organize and jump-start the growth of so many other processes.

Although there is enormous variability in the way newborns respond to being touched, there is remarkable consistency from baby to baby in terms of their bodies' physiological response. Gentle touching or vestibular stimulation such as rocking initially produces a state of biological arousal that resembles the classic stress response. Increased brain-stem activation triggers a series of chemical cascades involving all the usual suspects—increases in cortisol, ACTH, and other stress hormones. These effects reverse, however, with continued stimulation. During prolonged touch or slow, repetitive movements, the brain-stem systems that were once activated become inhibited, and slowly this process down-regulates the body's stress system. The stress hormones that were at first elevated begin to fall with continued stimulation and actually decrease below normal baseline levels (measured while the baby is at rest and not being touched or moved).

Studies now show that down-regulation of the stress system continues long after the touching or rocking stops. The effects of touch and motion are enduring, measurable hours after stimulation has ceased, and it is this property that seems to have such a significant impact on the development of other systems. Stress hormones such as cortisol have been found to have deleterious effects on synaptogenesis and synaptic pruning in laboratory animals, and brain limbic regions seem particularly susceptible. Rodents and primates that have been subjected to elevated cortisol levels show significantly reduced brain volume in several limbic regions including the hippocampus when compared to those that are not stressed. Moreover, animals that are given these experimental manipulations repeatedly for periods as short as two weeks exhibit a range of behavioral problems as juveniles that often persist into adulthood, including deficits in learning ability, memory, attention, sensory-motor integration, solving tasks that require flexibility and adaptability, and regulating their emotions. This suggests that the pleasure of touch and motion contributes to normal brain growth by regulating stress hormone levels during development.

These findings have led some researchers to ask if there are protective benefits for developing animals (including humans) exposed to stimulation and experiences that lower stress hormone levels. The answer, of course, is yes. For instance, gentle daily massage of laboratory mice for a period as short as two weeks during their first three months of life results in faster maturation of several cortical brain regions (hippocampus, somatosensory cortex, cerebellum) when compared to controls not given the stimulation. These mice are also less fearful and prefer novelty more than controls that are not given the extra stimulation.

The implications of studies like this have not been lost on neonatal care units. Preterm babies are now routinely swaddled in an effort to improve their rate of maturation and ensure that the prerequisite tactile sensations needed for proper brain development are experienced as much as possible. Some hospitals have even gone a step farther and offer swaddling in conjunction with programmed movement stimulation that occurs on specially constructed infant waterbeds. Although there have not been many controlled evaluations of these treatments, those that have been conducted have found that swaddling and movement therapy significantly increase the rate at which preterm babies reach critical developmental milestones when compared to preemies given the usual care. These benefits include reported increases in a number of behavioral measures such as learning to crawl and walk earlier, increases in formula intake and weight gain, improved responsiveness to touch, better muscle tone, increased visual acuity, recognition memory, and attention, among many others.

Older babies also benefit from touch and motion stimulation. In one experiment babies aged three to sixteen months were given scheduled vestibular stimulation by being placed on their parents' laps and rotated in a swivel chair for ten minutes each day over a period of two weeks. The babies, of course, loved the experiment—giggling and jumping during the rotation. Interestingly, the babies who were spun reached key developmental landmarks such as crawling and walking more quickly than their nonspun counterparts. This effect was even demonstrated in identical twins. The twin who received the motion stimulation began walking four months earlier than his brother.

Humans are programmed from birth to experience certain forms of touch and motion as intrinsically pleasurable. The roots of these hedonic preferences can be found in babies who are pacified by cradling and rocking; toddlers and children who self-stimulate their touch and vestibular systems using age-specific behaviors such as bouncing, rocking, and self-hugging; on through adolescents and young adults who have a need for speed. Like other hedonic preferences such as our desire for sweets, the pleasure we find in touch and motion satisfies critical developmental requirements for normal brain and behavioral maturation. The problem, however, is that technology has radically outpaced evolutionary pressures in the past two hundred years, leading to new and potentially harmful methods of satisfying this biological imperative. In chapter 11 we will consider how our need for touch and motion often couples with other hedonic preferences that together foster maladaptive behaviors such as addiction and thrill-seeking.

Chapter 5

In Praise of Odors

I will be arriving in Paris tomorrow evening. Don't wash.
—Napoleon to Josephine

All good kumrads you can tell by their altruistic smell.
—E. E. Cummings

No other sense is so intimately bound to memory and emotion as smell. To this day, the mere hint of something sulfuric takes me back to a steamy August birthday and the gift of a Junior Scientist Chemistry Set from my mother. The blurb on the back of the box was encouraging: "Perform over 1,500 experiments and procedures in the gaseous phases of matter, chemical models, solutions, acids, bases, electrochemistry, organic chemistry and more." For the next few weeks I couldn't stop playing with this thing. Mom was thrilled and, I'm sure, convinced that one day I would be the next Louis Pasteur. But parents often forget that eight-year-old boys are not terribly interested in analyzing the covalent bond properties of solvents or learning how to neutralize an acid; they tend to like things that make loud noises, blow up, or best of all, some combination of the two.

Scouring the list of experiments one evening, I found a promising entry called "Outrageous Ooze," which guaranteed an "explosive miniature volcanic reaction with real lava flow." Mom was busy cooking and cleaning in preparation for a dinner party at our house that night, and Dad was out driving across the state and back to pick up some fancy German chocolate ice cream for dessert. I was given advance warning to be on my best behavior, and yes, my friend Hector could come over as long as we played in the back room.

Hector was the best lab assistant I ever had; he was always eager to see what happened if we mixed this and that, and had a natural talent for combining compounds that we were warned against mixing. A rule for toy manufacturers: The phrase "Warning—never combine Chemical A and Chemical B" is usually translated by eight-year-olds into "Attention—please do this immediately." Just before the first guests began arriving I ran in the kitchen and asked my mother for some vinegar, baking soda, and dishwashing detergent. She hesitated for a moment, but then the doorbell rang and she didn't have time to protest.

Back in the lab, Hector and I mixed the ingredients carefully. We stood back and waited for the Vesuvial display, only to be disappointed by the slow trickle that emanated pathetically from the jar, so we consulted the next paragraph, which instructed us to "incorporate the following mixture to produce hydrogen sulfide gas for extra realism." I began heating the sulfuric mix before Hector finished reading the sentence, and suddenly we were treated to a thick display of smoke, and a horrible stench of rotten eggs began to fill the room. Even after removing the mixture from the heat, the pungent smell and smoke worsened and eventually, to my parents' horror, spread throughout the entire house. The memory I usually associate with this smell today consists of my parents and the complete dinner party standing out in the street watching fetid-smelling smoke billow from the front and side windows of our small house.

The chemical senses of smell and taste are as phylogentically ancient as touch, and their age is given away by basic anatomy. Whereas the perceptual seat of sight and sound resides in the neocortex—a recent addition in mammals—the representations of the chemical senses are stowed away in the rather archaic limbic and paralimbic regions. This is true for humans, primates, and virtually all mammals large and small.

The cortical representations of smell and taste are located in regions of the brain long believed to be important for processing the motivational state of an animal as well as the emotional significance of external stimuli. Experiments have shown that when humans are stimulated through taste or smell, large portions of the brain that are critical for processing emotional information and memory become activated, including the amygdala, insula, cingulate cortex, and orbitofrontal cortex.

Let's consider smell—it is the one sense that simply can't be turned off without immediate consequences. We can close our eyes, cover our ears, shut our mouths, and refrain from touching things, but stop breathing for a moment and you quickly realize that we are all slaves to olfaction. Humans take more than 23,000 breaths each day, passing close to 450 cubic feet of air through their nose. Our nasal passages act as miniature wind tunnels powered by a respiratory vacuum that induces air molecules to enter with astonishing force. Odor molecules have a bumpy ride as they enter the nose—first heated by the frictional forces as they pass on either side of the septum and then thrust up through three complicated horizontal chambers shaped by vascular tissue. The turbulent journey ends at the roof of these interior passages as the molecules collide with a small patch of yellowish tissue on either side of the septum known as the *olfactory epithelia*. At this point, the air molecules have reached the brain.

Each cell in the olfactory epithelia—and there are hundreds of thousands of them—has receptors that are tuned to a particular odor. The shape of the odor molecule is what matters most. If an odor molecule has a shape, or a very close match, that allows it to bind to one of the many olfactory epithelia cells, it can cause that cell to send a signal in the form of an action potential on to the next stage of neural processing. The sole job of the olfactory epithelia cells is to convert chemical signals that find their way up our noses into electrical signals that the brain will understand. Although we generally think of our sense of smell as being rather limited compared to other mammals— dogs, for example—humans can perceive and distinguish differences among thousands of odors.

In her book *A Natural History of the Senses*, Diane Ackerman refers to smell as "the mute sense." While we can detect and even perceive thousands of smells, we are woefully inept at describing them without reference to other things or, even more often, how they make us feel. This verbal shortfall may arise in part because the brain regions that register smells are only weakly and indirectly connected to those areas that support language processing. A more direct set of connections exists between areas that deal with emotions and language, and so the lexicon of smells is riddled with descriptions of how a smell makes us feel. Try to describe the smell of camphor without reference to a pine tree; or imagine explaining the smell of the ocean in the morning to someone who has never had the experience.

The history of olfaction is inextricably linked with the natural history of humans and the emergence of the first mammals. One theory suggests that during the Devonian period (about four hundred million years ago) life on Earth was dominated by aquatic species that used chemical senses to navigate their environment, find food, and attract mates. This may have taken the form of taste sensation or something similar, such as having appendages lined with receptor cells sensitive to the presence of amino acids. Nutritious food would have to be found by literally swimming through it. Many crustaceans still employ this form of chemical sampling.

A big improvement came with the appearance of the first nose, which was little more than a pair of epithelia pits or indentations on the early ancestors of the modern hagfish. These species had a significant advantage over competitors in that their primitive version of smell allowed them to detect food, mates, predators, and other elements important for their survival across extended distances. They no longer had to come into direct contact with an object to sense its presence; they only needed a sample of it in the form of volatile molecules unstable enough to diffuse through water or air.

The brain circuitry that processes olfactory information is essentially the same across all modern mammals. The differences are largely in terms of where the information is sent after reaching the primary olfactory cortex, and the sizes of the olfactory brain regions relative to other structures. For instance, rodents depend critically on a keen sense of smell, and their olfactory bulbs are enormous relative to other brain structures when compared to humans. This clearly has an impact on the ability of rodents to distinguish one smell from another, which is a key element of their survival. The basic mechanisms of olfactory sensation are the same as in humans, but not so heavily emphasized due to our equal reliance on the other senses.

Imagine you are walking before dinner one summer evening—on past the flowering dogwoods and myrtles that have exploded with color in the past few weeks, toward that unmistakable signature smell of the holiday weekend. You wonder if those are ribs or burgers, but after consulting your stomach decide that either would do. The chain of events that occurs between encountering the odor molecules and perceiving barbecued meat involves multiple stages of processing that provide a road map for understanding the evolution of smell in our species and the development of this sense in each individual.

The smell of barbecue is a complex mixture of scents. There is the smell that emanates from the charcoal, as well as from the cooking meat and flavorings. Each of these molecule types has different shapes and will activate different epithelia cells. The charcoal odorants will activate one set of epithelia cells, the cooking meat another set, and the smell of flavorings still other sets. Together, the group of activated cells forms an ensemble code that represents the complex barbecued meat smell that we actually perceive.

This signal is sent from the olfactory epithelia to the olfactory bulbs (one on each side of the brain), where it undergoes further processing and is then sent to several higher-level destinations. One copy is sent to the primary olfactory cortex, which is responsible for the conscious perception of the smell. A second copy is sent to the amygdala and adjacent structures that are responsible for translating motivational states such as hunger into appropriate responses such as feeding behaviors. Other copies are sent to limbic areas, including the hippocampus and the entorhinal cortex, which are critical for memory storage, as well as to the orbitofrontal cortex, which integrates the olfactory signals with those from other senses such as taste and assigns a reward value to the percept, in this case a hamburger. Hence, olfactory perception is situated in the primary olfactory cortex, and multisensory integration (for example, associating the smell of barbecue with taste information, which gives us the perception of flavor) with the reward value of a stimulus occurs in frontal locations that emerged later in our evolutionary lineage. Brain damage confined to the primary olfactory cortex—through stroke or physical trauma—leads to classic anosmia (an inability to smell and distinguish odors), while damage to the orbitofrontal cortex results in a complex syndrome of deficits in smell recognition and associated abilities that depend on multisensory integration.

When Melissa and I had our first glimpse of Kai at the sixth-week ultrasound, he was little more than a blastocyst, but even at this early stage in gestation he had the beginnings of an epithelia pit. From this point on in development, however, he shared fewer and fewer features in common with a hagfish embryo. At about eleven weeks into gestation, his olfactory epithelia cells began to extend toward cells that were beginning to grow in his olfactory bulb, and the bulb cells were, in turn, beginning to extend toward cortical sites. None of these developmental changes depends on smell experience, since until about the twenty-eighth week, Kai's nasal cavity will be filled with a soft tissue plug that prevents chemicals from stimulating these cells. Interestingly, olfactory epithelial and olfactory bulb cells do not reach biochemical maturity until about the twenty-sixth week into gestation, and this is precisely when they will begin to need stimulation to continue developing normally.

You may be inclined to think that fetuses probably can't smell very much, but research shows that their olfactory world is as rich as their mother's. By the twenty-eighth week, Kai's placenta has thinned to the point that virtually anything his mom smells is passed to him through the amniotic fluid. In fact, scientists have speculated that odor molecules may diffuse even faster in amniotic fluid than they do in air, since they ultimately must enter a liquid phase when binding to epithelia cells in the nasal mucus. So by the third trimester everything that Melissa eats and smells is experienced by Kai, and this has a huge impact on the continuing development of his nervous system and on olfactory preferences that will appear after his birth.

Once the nasal plugs are out and Kai begins to have his first encounters with smells, these experiences will kick the development of his olfactory system into overdrive, and the connections from the olfactory bulb to limbic and cortical brain regions will become more and more refined. First the connections between the olfactory bulb and limbic structures come online and allow Kai to perceive and distinguish among simple smells. These new connections allow Kai to perceive smells for the first time; however, the continued development of his olfactory system—most notably the important connections between the olfactory bulbs and higher cortical sites, such as the orbitofrontal cortex—depends critically on Kai receiving a wide variety of olfactory stimulation at this time, the more varied the better.

In animal models, if one of the two nasal passages remains sealed during this critical period so that no olfactory stimulation takes place, the corresponding epithelial cells, the olfactory bulb, and even cortical areas that normally would receive information from this side of the nose shrink up to 40 percent and lose cells rapidly. As expected, this results in a significant loss of smell perception and recognition after birth. Contrasting this, when premature animals born at thirty weeks are stimulated with an increased variety of smells (such as mint, cinnamon, banana, pine, or vanilla) through only one nasal passage, the olfactory brain regions that receive input from that side become larger and develop about 30 percent more cells than the control side that is stimulated with only ambient laboratory smells. Clearly, olfactory experience begins in the womb.

Animal experiments have also demonstrated that exposure to certain odors in utero has a dramatic influence on both pre- and postnatal behaviors. Rat fetuses display a sudden increase in excitable activity after pleasurable scents such as mint or lemon are injected into the amniotic fluid. Injections of simple saline solution or comparably bland scents have no apparent effects. After birth, the rats that were exposed to a mint or lemon scent while still in the womb prefer to nurse on nipples where these scents are present, rather than on those with neutral scents, a behavioral preference that keeps the pups near odors associated with the maternal environment.

Rats can also be classically conditioned to odors while in the womb. If their amniotic fluid is scented with an odor (even a pleasurable odor such as apple) and the fetus is then injected with a substance that makes it nauseous, it will avoid places and objects that bear that scent after birth. Such conditioned taste aversion was once thought to occur only in more mature animals, but it is now clear that prenatal animals are capable of many forms of learning.

These data tell us three very important things about olfaction. First, fetuses have a significant capacity for olfactory learning, since they remember a scent associated with the womb and seek it out after birth. Second, certain odors are innately excitable or pleasurable to animals in that they can function as primary reinforcers of behavior and have an impact on behavior and physiological responses the very first time they are experienced. Finally, the capacity for olfactory learning and memory can offset innate odor preferences, making a scent that is normally attractive something to avoid after birth.

Humans show remarkably similar forms of olfactory learning, and prenatal exposure to odors seems to play an important role in parental bonding and kin recognition. Newborns have an innate fondness for the smell of amniotic fluid, particularly their own. Experiments performed in culturally diverse populations have shown that babies as young as one day old prefer the smell of their own amniotic fluid to that of age-matched controls. The most commonly used test of preference in newborns is, of course, sucking behavior. When given the choice between nursing on their mother's breast scented with their amniotic fluid or that of an age-matched control, they almost always choose the former. Newborns also cry less and show a diminished stress response when they smell their own amniotic fluid. Since the many odors that emanate from a mother—such as the smell of milk, colostrum, saliva, and perspiration—stem from the same genetic and dietary sources as the amniotic fluid, they will all have some shared chemical groups. Hence, a preference for the smell of maternal amniotic fluid may evolve functionally into a preference for the smells of Mom in general; an adaptation such as this would have obvious utility in keeping the newborn close to its primary caregiver. These behaviors are evidence that olfactory labeling occurs in humans as it does in other animals.

Olfactory labeling while still in the womb has a profound influence on our postnatal ability to identify a person as kith and kin. Within hours after birth, a breast-fed infant can readily identify and will orient toward a breast pad worn by their lactating mother over a breast pad worn by an unrelated lactating woman. Newborns show abrupt changes in behavior—such as decreasing arm and leg movements, initiation of the sucking reflex, and are generally calmer—when exposed to odors that originate from their mother's body, including those that emanate from her breast, underarms, and neck. Almost any natural smell that can be used as a reliable indicator of Mom's presence has these effects on behavior.

Olfactory labeling also has an impact on the development of odor preferences and aversions after birth. In the Alsatian region of France, there is widespread use of anise flavoring in the local cuisine. Taking advantage of this custom, a group of scientists compared the olfactory responsiveness of neonates who were born to mothers who had or had not consumed anise flavor during pregnancy. Infants born to anise-consuming mothers showed a stable preference for the smell of anise when tested immediately after birth and four days later.

Contrasting this pattern, infants born to mothers who did not consume anise tended to display either an aversion or no response at all to the odor. This study indicates that olfactory labeling also occurs in response to dietary influences that may alter the in utero chemical environment. One can imagine the profound implications of this process for the newborn in that it most likely influences a host of functions that range from the emergence of odor and food preferences to the development of early mother-infant attachment.

Although not all scientists agree, it appears that newborns may also have innate preferences for certain odors, such as floral and fruity smells. These odorants are not necessarily present in the amnion of most mothers, yet the preference for these smells emerges in most newborns across different cultures and persists into childhood. Researchers have found that newborns can discriminate among a number of qualitatively different odorants, evidenced by changes in body movements, facial responses, and heart and respiratory rates. It is much easier, however, to test verbal children who can simply tell you whether they find a smell pleasing. The few published studies that have focused on the olfactory preferences of verbal children have not always found consistent effects, and one reason may be that differences in experimental design influence the results. For instance, it is well known that young children tend to answer a positively phrased question in the affirmative. When these and similar methodological issues are controlled for, however, some universal tendencies do indeed emerge.

It is generally accepted that children as young as three years old exhibit stable hedonic preferences for specific odors independent of the culture in which they were raised. Some of the most popular smells include strawberry, floral, spearmint, and wintergreen, while odors such as butyric acid (strong cheese/vomit) and pyridine (spoiled milk) are universally disliked. That most children find fruit and floral odors pleasing should come as no surprise, since they often signal the presence of a nearby nutritionally rich food source—an important adaptation, to be sure, within an evolutionary context.

The emergence of this evolutionary adaptation in our species is echoed in the development of each individual with the growth and maturity of brain pathways that connect olfactory cortical areas with midbrain and orbitofrontal regions that mediate natural reward. While the midbrain reward centers develop at a fairly early embryonic stage in humans, the pathways that connect these regions to the areas responsible for perceiving odors do not mature until rather late in gestation and are known to depend on experience. Because this is the case, many scientists believe that although these olfactory preferences are very similar across cultures, their development probably results from learning to associate these smells (in the womb) with flavorful food and the onset of an intrinsically rewarding behavior—eating.

Consummatory behaviors such as eating and sexual activity are known to increase levels of circulating neuropeptides called endorphins, those lovely chemicals that provide a feeling of relaxation and calm and that are chemically similar to morphine. The development of these pathways—which depend on exposure to odors that signal the presence of potential sources of nutrition—may occur during the last trimester, while the fetus is exposed to the coappearance of certain smells and tastes with an increase in amniotic endorphin levels that have a calming effect on mother and fetus alike. Hence, the emergence of "universal" olfactory preferences is likely to result from the same learning mechanisms that mediate olfactory labeling in all mammals.

The Smell of Attraction

Like it or not, we smell, and the subtle odorous messages we send and receive—often unknowingly—have a profound influence on our social identities and a wide range of behaviors, including mate selection, courtship, and the timing of ovulation. The word *pheromone* calls up a variety of images to mind: mammals communicating using a hidden language of scents; trendy socialites paying $300 per ounce for a vial of boar effluvia that promises to allure the opposite sex; and sorority sisters who menstruate in synchrony month after month.

Although it has proven rather difficult to isolate and identify a human pheromone, there is a growing body of evidence that we use them to communicate chemically much like other mammals. The first convincing evidence came from an unexpected place—an undergraduate dormitory room at Wellesley College. In 1967, an undergraduate student named Martha McClintock noticed that many of the girls in her dorm menstruated on the same days and wondered if such coordination might have survival value. She asked two simple questions in her research project: "When did you last menstruate?" and "Who are your two best friends?" The results surprised everyone. Women who spent the most time together tended to menstruate at the same time.

It wasn't until ten years later that the mechanism that causes this synchrony (now known as the McClintock Effect) was discovered. In a simple experiment, psychologist Michael Russell and his colleagues at Sonoma State Hospital in California rubbed an extract from the underarm of a woman with a very regular twenty-eight-day cycle under the noses of sixteen other women three times a week. Within four months, all of the women were menstruating within three days of one another. The odors of a single person, it turns out, can influence the menstrual cycles of many others. It was still unclear, however, how such an effect might have survival or adaptive value.

The answer came the following year, when it was discovered that men have cycles as well, and that their regular rise and fall in core body temperature and the production of essential steroids such as testosterone can be modulated by the presence of other males. The final link came when additional experiments showed that the production of testosterone and other androgens in men often becomes synchronized to the menstrual cycles of their wives and lovers. Taken together, an impressive display of synchrony emerges between men and women in close contact with each other on a regular basis, and this patterning might facilitate the timing needed for effective sexual reproduction.

In addition to regulating menstrual and physiological cycles, pheromones have a say in other very personal affairs, such as distinguishing those we find sexually attractive from those who remind us of a sibling. But what is the physical basis for this hidden conversation of scents? Humans don't seem very interested in smelling the urine or underarm odor of potential mates, so where do human pheromones come from and what do they smell like?

The Desana Indians of the Amazon rain forest have a cosmological worldview built around olfaction. For them, the essence of a person is revealed by their smell, which emanates directly from their bones. Mores that guide courtship and social relationships are intertwined with the relationships among different smells. The Desana believe that people of the same tribe share a common smell, and strict rules exist forbidding marriage between those who have similar scents—an olfactory-inspired incest taboo. Certain smells should never be mixed, yet some naturally go together. An answer to the puzzle of how human pheromones influence behaviors such as mate selection came out of left field and was inspired by studies of the Desana.

It has been known for ages that foreign tissue implanted into a host is often rejected by the host's immune system. Each of our own cells bears proteins that our immune system recognizes as "self." When a foreign cell enters the body, the immune system attempts to classify the nonself intruder by attaching a labeling protein to it and generating antibodies designed to destroy it. The immune system has a memory for intruder cells, and the next time the same foreigner is encountered, the antibodies can be launched even faster, since they do not need to be generated from scratch.

A segment of our DNA called the major histocompatability complex (MHC) codes for the immune cells that identify intruding disease organisms, essentially functioning as our immune system's first line of defense. Unlike many genes that have only a few alternative versions (called alleles), MHC genes have upward of a hundred or so, with each providing immunity against different sets of potential disease strains.

When we think of heredity, we typically have in mind the classical pattern of single allele combinations—the dominant-recessive pairings that play a winner-take-all game with traits such as eye and hair color. If one parent has blue eyes and the other brown, one gene will dominate, meaning the gene from the other parent that controls this trait is not expressed.

MHC genes work differently in that they are codominant. Say you inherit one version of an MHC gene from your father that improves resistance to disease A and another version of the gene from your mother that happens to help fight disease B. Since MHC genes are codominant, you will be able to resist both diseases. Thus, parental combinations that have the greatest degree of MHC genetic heterozygosity* will produce offspring with the most robust immune functioning. These offspring would have a distinct survival advantage over offspring from parents who have considerable overlap in their MHC genes, providing resistance to a smaller spectrum of disease strains. Variation in this dimension, then, can serve as an important selection factor in the evolution of our species. The question is, what serves to attract us toward mates who have MHC genes different from our own?

The initial clues emerged from animal experiments. If a female mouse is offered two suitors, she inevitably chooses the mate whose MHC genes have the least overlap with her own, and it is now known that they do this through smell. Scientists have found that each version of the MHC gene codes protein by-products that are excreted from the body, and they have a unique odor. Mice that have damage to their olfactory nerve or olfactory epithelia cells perform this test at chance levels—deciding on the heterozygous mate only about half of the time.

So if you're a smart rodent, a big part of your mate selection process is in deciding if a suitor has the right smell. Mice that are most attracted to the smells of potential mates with dissimilar versions of MHC genes will be less likely to inbreed and will maximize the genetic fitness of their offspring. Can such a process be important for human mate selection? The fact that perfume and cologne sales account for approximately 12 to 15 percent of annual consumer

---

*"Heterozygosity" means having dissimilar pairs of genes for any hereditary characteristic.

luxury item spending suggests that we believe smell is a key factor in shaping our own attractiveness.

Intrigued by the idea that humans may use a very similar process to select mates, evolutionary psychologist Chris Wedekind and his colleagues conducted an experiment in which they asked more than one hundred men and women to score the odor of T-shirts worn for two consecutive days by male and female subjects. Each person tested was brought into a room with six odorous T-shirts stored in separate plastic containers and asked to rank them in terms of their "sexiness" and "pleasantness." The results were surprising.

Scores of the pleasantness and sexiness were indeed found to relate to the degree of MHC similarity between the smeller and the T-shirt wearer. For most subjects, the most pleasant and sexy smells were associated with members of the opposite sex whose MHC genes had the least overlap with their own. When asked why they liked a specific smell, many subjects offered that it reminded them of their present lover or an ex-mate. Interestingly, lower-ranking odors were said to remind the smeller of a sibling or other relative. This is one case where opposites definitely attract, and for a good reason— the observed mating preferences stemming from these choices would naturally increase immune system heterozygosity of the offspring.

In this mechanism we find that there is no single attractive smell that works for everyone—one person's *sentir bon* may repulse another. To take advantage of this adaptation, lonely singles in search of a mate would have to get a genetic fingerprint of his or her intended before a scent could be custom-designed. But all hope is not lost. In 2001, Wedekind's group showed that we unwittingly use our own MHC geneotype information in choosing perfume and cologne for personal use. In this study 137 male and female students who had been typed for MHC were asked to rate 36 different scents for personal use—"Would you like to smell like that yourself?" The researchers found a significant correlation between MHC genotype and scent rankings, indicating that people with similar MHC alleles preferred to wear similar-smelling perfumes. These results suggest "that perfumes are selected 'for self' to amplify in some way body odors that reveal a person's immunogenetics." While it is commonly assumed that perfumes are worn primarily to mask a person's natural odors, Wedekind has argued that we actually prefer to wear scents that accentuate these olfactory cues, announcing our MHC genotype through a form of olfactory advertising.

In this chapter we've seen that there are a few treasured scents that are universally appealing. Floral and fruity smells top the list in most countries, probably because they signify the presence of nutritious food sources. Newborns and children alike are attracted to these scents, independent of the culture in which they were raised.

Newborns and infants are also universally attracted to the smell of their own amniotic fluid and those that are breast-fed come to associate maternal odors with the arrival of food. Attraction to maternal odors has obvious survival benefits by keeping the offspring in close proximity to its mother.

We have been wired by natural selection factors to find pleasure in these and similar smells because they have survival value. Pheromones, on the other hand, have a different kind of universal appeal. While no single odor is pleasurable to everyone, the rule is very simple (and universal): sexy/pleasurable smells signify the presence of potential mates that can lead to viable offspring. The pleasure we find in these "hidden" scents is driven not by natural selection factors, but rather through sexual selection because these adaptations have clear reproductive value.

In the next chapter we turn our search for pleasure toward the epicurean in us all. We will discover that our lust for certain tastes fosters normal immune system and brain development, but at a growing cost to public health in Western societies.

Chapter 6

For the Love of Chocolate

*As life's pleasures go, food is second only to sex.
Except for salami and eggs. Now that's better than
sex, but only if the salami is thickly sliced.*

—Comedian Alan King

*Music with dinner is an insult both to the cook and
the violinist.*

—G. K. Chesterton

If you ever make it to the Amazon, ask your guide to show you what is likely the most revered tree throughout all of South and Central America. The *theobromo cacao*, or "food of the gods," was named by Linnaeus, the great eighteenth-century cataloger of nature. The designation makes clear his admiration for the almost indescribably savory taste of its fruit and seeds as well as the role of the tree in world history. Before taking a bite, few people stop to think—and who can blame them?—about the influence of chocolate on the social, political, and economic evolution of those cultures that came in contact with cacao beans as they spread from equatorial America to Europe, and then Asia.

After a short trek into the jungle, your guide will stop in front of an odd-looking tree, probably no taller than ten meters or so. If it is a mature tree—older than three years—it will have large patches of pink or blue cauliflorous growth on its bark, but your eyes will skip right past this feature and focus on the strange football-size pods that dangle expectantly from its trunk. The outer covering of the tree's fruit is a tough hide of corrugated green and yellow, which when broken reveals a soft, whitish pulp. The taste of the pulp will catch you off guard. Most people anticipate the tangy sweet flavor of fruit with their first bite, only to be surprised by the subtle, bittersweet taste of chocolate. Enveloped inside the pulp are dark, purple-colored seeds—about thirty to forty per pod—that after being dried and processed can be recognized by epicureans around the world as "chocolate beans."

Few foods inspire such passion in people as chocolate. This love affair goes far beyond the typical fondness for sweets: after all, we're not likely to head out into a snowy night, panic-stricken after finding that we are out of lemon-crème pie or bubble gum. There is something special about chocolate that drives us to extraordinary lengths. Chocoholics find nothing strange in spending a small fortune for even a sampler box of champagne truffles. No, a simple sweet addiction is not the same as a chocolate addiction—indeed, many connoisseurs prefer the darkest, most bitter variety.

The history of the chocolate bean is a story riddled with desire that transcends cultural distinctions. The roots of chocolate go back some twenty-six hundred years to the great Olmec and Mayan civilizations that flourished throughout southern Mexico, Belize, Guatemala, and Honduras. Spouted, teapot-shaped vessels have been excavated from towns such as Colha in northern Belize, and found to contain residue of ancient chocolate. The Mayan drink was very different from the watery, sugar-laden version of hot chocolate that dominates modern society. The journals of Spanish conquistadors are filled with descriptions of middle Mayan culture that include the preparation of dried cacao beans ground into a powder and mixed with water, honey, chili pepper, and sometimes maize. The liquid would then be heated and repeatedly poured from one vessel to another to produce a thick head of rich chocolate foam that was the most coveted part of the drink.

A reverence for chocolate was also present in Aztec culture throughout the region. The great Aztec emperor Moctezuma reportedly drank up to fifty flagons of chocolate per day, believing it to have restorative and even aphrodisiac powers. Within this culture, the cacao bean became the primary form of currency, and folklore has it that when the Spanish conquistadors stormed Moctezuma's temple, they found beans in place of gold.

After conquering the Aztecs, Hernando Cortés returned to Spain and brought King Carlos treasures of cacao beans and a recipe for making *xocoatl*, which was sweetened with sugar by members of his court. Today this recipe lives on, and the domesticated cacao tree grows on farmlands near the equator in a number of regions including the Caribbean, Africa, southeastern Asia, and in several South Pacific islands such as Samoa and New Guinea.

Whereas modern processing and distribution technology has made chocolate a more common item on the food landscape, its seductive properties have remained a mystery to science. It has only been in the past few years that neuroscientists and biochemists have begun to get a handle on why we find chocolate so pleasurable.

Chocolate contains more than 350 known compounds, several of which activate three important brain systems that contribute to the experience of pleasure. The first ingredient that gives chocolate its wide fan base is plain old sugar, an underappreciated compound these days. Considering our modern tendency to detest all that is carbohydrate and the epidemic-like rates of diabetes in many subpopulations, it is easy to understand why sugar is seen as something to avoid if you consider yourself a health-conscious individual. But in reasonable doses, sugars have a profound and positive impact on our physiology, most notably in the form of a calming effect. Placing a small amount of liquid sweetened with either glucose or sucrose on the tongue of a crying newborn has an immediate calming effect that can last for several minutes. Sugars, in their varying chemical structures from lactose to sucrose, have been shown to activate the brain's opioid system, a set of circuitry that plays a prominent role in regulating the body's stress response.

In addition to sucrose, chocolate contains small amounts of theobromine (a mild stimulant) and phenylethylamine, a substance that is chemically similar to amphetamine. Once in the brain, each of these ingredients has an effect on the dopamine and noradrenergic neurotransmitter systems, which are implicated in attention and general arousal. These compounds are thought to provide the "boost" we all experience after eating chocolate.

But chocolate gives us more than a mere boost; most people crave the sense of euphoria that lingers long after the treat is gone. A recently discovered trio of chemicals has been identified in chocolate that seems to be at the heart of this feeling of well-being that is familiar to all chocoholics. Anandamide is a chemical messenger in the brain that binds to the same nerve cell receptors that are activated by tetrahydrocannabinol (THC)—that's right, the active compound in marijuana. Anandamide, it turns out, is released in small quantities during times of stress and provides a calming and analgesic effect; however, it is quickly broken down by naturally produced enzymes, so there is never very much of the substance in the brain under normal circumstances. The buzz one gets from marijuana is another story altogether—in this case a deluge of THC enters the brain, overwhelming the ability of the enzymes to break it down, so it has a prolonged and more intense effect than the naturally occurring version. The "THC buzz" is essentially an exaggeration or amplification of normal cannibinoid brain system functioning.

The chocolate buzz occurs through a slightly different mechanism. Small amounts of anandamide are present in chocolate (darker chocolates have larger quantities), but not so much that would activate the brain's cannibinoid system above normal. The key to unraveling this mystery came when two additional anandamide-like compounds were identified in chocolate and found to be present in fairly large quantities. While these related compounds don't activate THC receptors directly, they increase the effect of naturally occurring anandamide by blocking the enzymes that usually break it down. This means that even small amounts of naturally occurring anandamide or that ingested while eating chocolate will stay in the brain for a prolonged period of time, since it is not metabolized as quickly as normal. The result is that blissed-out feeling of calm that we experience after downing a hot chocolate or going through a few Droste pastilles.

It is easy to see why the depressed and stressed among us self-medicate with chocolate. It quenches the pleasure instinct by activating three key brain transmitter systems that are involved in reward, although they have evolved as adaptations to very different environmental circumstances.

The sucrose in chocolate is just a "souped-up" version of fructose—a form of sugar that is naturally present in most fruits that were widely available to early hominid hunter-gatherers. Sugars are a critical component of life because they provide metabolic energy in the form of ATP that powers the many biochemical reactions within every cell of our body. For the average hunter-gatherer, fruits were a very good nutritional choice, since ounce for ounce they offer a rich source of energy with virtually no exposure to dangerous horns, teeth, or claws. The only problem is in identifying fruits as a desirable substance to eat.

Early in our hominid evolution, the brain opioid system became very important for controlling our eating behavior, mainly in functioning to make sure that certain foods seemed more palatable than others. During this point in the evolutionary history of humans, opioid system activation and the pleasurable sensations that result became associated with the consumption of foods that have relatively high concentrations of sugar. This association was strictly in terms of alterations in brain wiring—some hominids evolved changes in their opioid system that made its indirect activation possible through receptors that were sensitive to the presence of sugar. In a very real sense, the opioid system, which until then probably played a role mainly in sexual reproduction, was co-opted by selection factors that made it very cost-effective (energywise) for hominids to be able to find and want to eat sugar-rich fruits. Hominids with a tendency to experience pleasure when eating something sweetened by natural sugar had a clear survival advantage over their peers who were not "afflicted" with this important mutation to opioid system wiring. Similar mutations may have occurred within the dopamine reward system and the cannibinoid system, making the taste of chocolate a "triple threat."

Sugar and Health

Modern societies have very different survival pressures, and hence selection factors, than those of early hominids. Not only do we have access to all the fruits we want, we have also perfected the packaging and delivery of refined sugar such as sucrose in a staggering variety of forms that include processed foods and candies. Our opioid systems are awash in a sea of sweet-tasting stimulants, and this has serious consequences for societal health.

Study after study has shown that in humans, the most palatable foods release the highest levels of beta-endorphins into our blood-stream. Endorphins are the brain's natural opioids that are typically released during stress. When this system is rendered inactive by administering an opioid antagonist (drugs that bind to opioid receptors but do not activate them) such as naltrexone or naloxone, subjects report that foods taste less palatable and food consumption is often substantially reduced. Thus there is very clear evidence that the opioid system is involved in the hedonic experience of food.

In the past decade research has shown that obese people often have a different opioid system response compared with nonobese individuals. At least two independent studies have found that obese subjects produce up to three times as much beta-endorphin in their blood plasma after consuming a palatable meal when compared to their skinnier counterparts. One interpretation of this finding is that some individuals may be predisposed to obesity because they have hyperactivated opioid systems and literally experience more intense pleasure in response to opioid-system-activating foods than others. It is currently unknown, however, whether this change in opioid system functioning is a cause or a result of obesity.

The opioid system also plays an important role in attachment behaviors. In mouse pups, the response to being separated from their mother consists of ultrasonic vocalizations accompanied by a brief period of hyperactivity until the two are reunited. Mice that have their opioid system blocked by chemical agents or through genetic manipulations fail to display the same plaintive calls as normal mice. They do, however, protest to other events such as sudden changes in temperature or the introduction of an adult male. Hence, attachment behaviors depend on opioid system activation.

The fact that attachment behaviors seem to involve the opioid system is interesting when one considers that breast milk is rich in lactose, a sugar that serves to stimulate the activation of this system. Newborns and infants are innately attracted to the smell (see chapter 5) and taste of breast milk, and they have an uncanny ability to identify milk from their own mother. By the end of the first week of life, newborns prefer the taste of their mother's breast milk over cow's milk. There are many compounds in mother's milk that may account for this preference. Besides being sweetened with the sugar lactose, it is rich in essential fatty acids that we will see are sought out by virtually all humans, from newborns to adults. Additionally, mother's milk contains a number of important immune factors and whole immune cells (one reason that synthesizing human breast milk has not been possible for manufacturers of baby formula) that may be critical in helping the infant identify and develop a preference for its own mother's milk over milk from an unrelated lactating mother.

Hence, newborns that are allowed to breast-feed will naturally self-stimulate their own opioid system, which itself may be a necessary component for the development of normal attachment. The timing of this sequence of behaviors is important, since the ingestion of lactose occurs coincidentally with other forms of stimulation that are known to activate the opioid system, such as the sensation of being touched and the familiar smell of Mom. All of these stimuli activate the opioid system at a highly opportune time for the development of maternal-offspring attachment—during feeding behavior. A biological mechanism such as this, which uses the experience of pleasure to prod newborns toward behaviors that at once maximize both attachment and the intake of nutrition, is likely to have tremendous survival value.

Getting Wired for Taste

Modern science has identified five basic food tastes: sweet, sour, bitter, salty, and the latest entry, umami, which is caused by the presence of monosodium glutamate (MSG). The first step in the path toward taste perception begins with the humble taste bud. Under an electron microscope, a taste bud has a shrublike appearance—think rhododendron—that is shaped by forty or so elongated epithelial cells. Each epithelial cell has receptors that preferentially respond to the presence of compounds affiliated with one of the five taste groups. As I eat my lunch of stir-fried vegetables, the natural sugars in the carrots and snow peas that pass by the approximately five thousand taste buds that line the perimeter of my tongue will activate groups of cells that are most sensitive to sweet-tasting food; salt-sensitive cells will become activated in response to the presence of sodium and potassium in the veggies and sauce added for seasoning; still other cells will be excited by the MSG.

The collective ensemble of activated cells sends this taste information on to the next stage of processing in the medulla and other nearby brain-stem structures that control the automatic behaviors involved in feeding such as sucking, salivation, and swallowing. From the brain stem, this signal makes its way to the thalamus, and finally branches out to cortical gustatory areas where the conscious perception of taste occurs and to various limbic nuclei where taste information can be integrated with memory, emotions, and motivation centers that regulate our desire to eat.

From the basic physiology and anatomy of the two chemical senses, we know that taste and smell are processed in very different ways. Epithelial cells in the olfactory mucosa respond to thousands of different types of odorants, while those that make up the gustatory system seem to have evolved a preference for basically five dominant taste classes. Besides this difference, however, there are remarkable similarities in the evolution of these systems in the human species and their development in the individual.

By the beginning of the second trimester of pregnancy—just about the time when Melissa was regaining her desire to eat and morning sickness was making a thankful exit—Kai's taste buds were beginning to mature. It is probably no coincidence that his first real sucking and swallowing behaviors also started at about this time, since the continued development of his taste buds, and most importantly their anatomical connection into functional taste circuitry in the brain-stem, depend on stimulation. The brain-stem sites mature very early as well and will provide Kai with a complete set of reflexive movement patterns for getting the nutrition he needs—everything from sucking and swallowing behavior to changes in facial expression in response to sweet versus bitter tastes. But it is unlikely that fetuses can consciously perceive tastes at this point in their development, since cortical taste sites are not yet mature. Anencephalic newborns lack most of their cerebral cortex, yet they are capable of the same behaviors. These include tongue protrusions to reject bitter-tasting liquids, and salivation in response to sweets, even though detailed investigations show that these infants have no genuine awareness of such tastes.

Comparisons across a wide range of mammalian species has shown that the taste circuitry that projects from the epithelial cells to brain-stem sites is highly conserved across very different animals, and hence is likely to have evolved rather early in the evolutionary lineage of hominids. Just as the conscious perception of taste and its integration with brain systems that regulate pleasure are likely to be relatively newer adaptations built on existing brain-stem circuitry, so it is for the developing fetus, who fails to show signs of real taste preferences until about the third trimester, when brain-stem connections to cortical and limbic regions are complete.

At this point in his development, Kai is experiencing all sorts of tastes—sweets, sours, bitters, you name it—all of which are incorporated into the amniotic fluid through Mom's diet. Even before my son is born, he has a sweet tooth. Although he tends to move most in the late evening hours, his fetal gymnastics can be brought on at any time of the day if his mother indulges in a bowl of Häagen-Dazs's Dulce de Leche. And this is not unusual. Before the days of ultrasound, X-ray contrasts were commonly used to assess fetus health in the final trimester. Studies performed during this time show that fetuses increase their swallowing behavior and movements if a sweet solution such as saccharine is injected into the amniotic fluid, while they decrease their swallowing if a bitter or noxious-tasting substance is injected. These results are consistent with the idea that taste perception and preferences emerge during this developmental period.

Well before Kai has any exposure to the outside world, he is already establishing taste preferences that will form a lifetime of eating habits. Evidence from both animal and human research indicates that taste variety is remarkably important during this stage of development. For instance, rats born to mothers who have had their salt intake curtailed during the final stages of gestation lack the ability to perceive the substance after birth. Likewise, rats born to mothers who consume diets rich in particular tastes such as apple juice or alcohol show an enhanced preference for the taste after birth when compared to rats born from mothers with a normal diet. Both of these forms of experience-expectant learning also occur in humans.

Finally, it is important to note that a very general relationship appears to exist between the experience of taste variety in the womb and acceptance of novel foods after birth. Newborn rats and humans exposed to an increased variety of tastes in utero typically show less fear of novel foods when compared to newborns from mothers who had a more restricted diet.

Much like we saw with smell, although newborns have an innate preference for specific tastes such as sweets and certain fats, they also exhibit an impressive potential for developing novel taste preferences based on what was experienced in the womb. Moreover, these experiments demonstrate that the development of normal taste perception depends critically on experiencing a wide variety of tastes while in the womb, since limited exposure to a taste class (for example, salts) can result in a reduced ability to detect and perceive these tastes after birth.

Survival of the Fattest

So far we've seen that humans find the consumption of sweets innately pleasurable, and that the evolution of this tendency can be traced to the evolutionary pressure to identify and *desire* high-energy foods (such as fruits and mother's milk) that are rich in natural sugars and relatively plentiful and safe to consume. But what about fats? Why do humans have such an insatiable appetite for fatty foods?

Although many of the ancient Greeks, including Aristotle, considered fat a basic taste class, it has only been in the past few years that food scientists and psychologists are willing to accept the idea that fat has a specific taste. Previously, most scientists believed that fat only acted as a food texture or flavor carrier. But this has changed with the discovery that simply putting a fatty food such as cream cheese into your mouth raises blood serum levels of triacylglycerol (TAG), an indicator of blood fat loading, even if the food is never swallowed. Richard Mattes, a food scientist at Purdue University, and his students followed up on this original study by showing that blocking the

subjects' ability to smell the cream cheese has no effect on the outcome, suggesting that it is the taste component of a fat that produces
this change in blood TAG levels.

These findings are probably no surprise to researchers such as
physiologist Adam Drewnowski, who in the early 1980s showed
that subjects' rating of the pleasantness of a food is directly related to
the relative proportions of sucrose and fat in the samples tested. We
all love foods that are laden with sugar, but there is a limit beyond
which we find a food to be too sweet. Hedonic preference ratings
first rise and then typically decline with increasing sucrose concentration in these experimental studies. This is not, however, what
happens with fatty foods. Surprisingly, hedonic preference ratings
typically continue to rise with increases in dietary fat content. It is
possible, then, that our innate fondness for fats is even more intense
than for sweets. And this makes perfect sense from both evolutionary
and developmental perspectives.

Let's start with evolution. In the past 2.5 million years, the hominid
lineage leading to humans has evolved significantly larger brains relative to body size when compared to other primates. Understanding
the reason for this dramatic expansion has been a long-standing
question for those concerned with human evolution. Many theories
argue that brain expansion followed the development of some key
cognitive or behavioral milestone such as the emergence of bipedalism or language or social group formation or toolmaking, and so on.
The list is long and varied, but the question remains: Did these new
functional capacities result from or cause the dramatic increase in
hominid brain size?

Michael Crawford of the Institute for Brain Chemistry and Human
Nutrition in London has argued that hominid brain expansion is
the direct result of dietary shifts that accompanied the migration
of *Homo sapiens* from the open savannas to freshwater and saltwater
shoreline regions, predominantly in the East African Rift Valley some
250,000 years ago. Human babies have combined brain and body fat
that accounts for a whopping 22 to 28 percent of their total body
weight, a finding that is not seen in any other terrestrial animals. Fats

are an indispensable component for building brains. The very foundation of life—the cell membrane—is made from a double layer of lipids that protects and shields the internal organs of the cell while at the same time permitting the perfect amount of elasticity so the cell can respond to physical changes in the extracellular environment. In human babies, a high level of dietary fat is critical for normal brain development because it provides energy for growth in the form of fatty acids found in triglycerides; contains important chemical precursors to ketone bodies that regulate brain lipid synthesis; and provides a store of long-chain polyunsaturated fatty acids, most notably docosahexaenoic acid (DHA) and arachidonic acid (AA), which are essential for the formation of retinas and synaptic junctions where brain cells communicate.

Both DHA and AA are present in abundance in human milk but noticeably absent in cow's milk. Recognizing the importance of these fatty acids for human brain growth and development, many formula manufacturers have begun supplementing their existing recipes with DHA and AA. Human body fat contains more DHA and AA at birth than at any other time during life, and in the newborn approximately 75 percent of its total energy expenditure goes to brain growth. Hence the fatty acids DHA and AA are important for brain development because they serve as an energy supply to fuel cell growth and proliferation, and because they have a molecular structure that is a unique component for building synapses.

Michael Crawford and his colleagues have suggested that since the natural supply of DHA and AA in human newborns is only enough to last the first three months of life or so, the continued supply of these fatty acids must occur through the child's diet. This means that the availability of foods that are natural sources of DHA and AA is a rate-limiting factor on human brain development and would naturally restrict the expansion of hominid brain size throughout evolutionary history. So where do you find rich veins of DHA and AA? Both are part of the larger omega family of fats whose synthesis requires the presence of two essential fatty acids that are not manufactured by the body, and consequently must be obtained through the diet. Alpha-linolenic acid (ALA) is the foundation of the omega-3 family of fatty acids that your body uses to make DHA, and linoleic acid (LA) is the foundation of the omega-6 family that is used to make AA. Both substances emerged in response to evolutionary pressures in plants to efficiently store and access energy reserves. Photoplankton, algae, and green leaves synthesize ALA in their chloroplasts, while flowering, seed-bearing plants store lipids in the form of seed oils loaded with LA.

Crawford's group has argued that the evolution of the hominid brain to the human form we know today would have been impossible unless early *Homo sapiens* incorporated large amounts of both ALA and LA into their daily diet. They suggest that the most dramatic increase in hominid brain expansion co-occurred with the migration of *Homo sapiens* to shoreline environments and lacustrine estuaries, where dietary ALA and LA were plentiful.

Whether or not Crawford's hypothesis is correct, two things are absolutely clear: growing brains require significant amounts of both ALA and LA, and these critical ingredients must be obtained through diet. ALA and LA deficiency in animals and humans results in altered structure and function of brain cell membranes and can lead to severe cerebral abnormalities. These anatomical changes have been linked to a number of disorders. For instance, both ALA and LA are involved in the prevention of some aspects of cardiovascular disease (including cerebral vascularization), and reduced levels of the fatty acids have recently been attributed as a cause of stroke, visual deficits, and several neuropsychiatric disorders, including depression, presenile dementia, and most notably Alzheimer's disease. Another study showed that ALA deficiency decreases the perception of pleasure by directly altering the efficacy of sensory organs and by creating abnormal changes in frontal cortex anatomy.

Taken together, these findings provide compelling evidence that many selection factors were operating to ensure that early *Homo sapiens* with a taste for foods containing ALA or LA would have a survival advantage over their peers who lacked this preference. Nature has solved this problem by giving us not just a "sweet tooth," but also an appetite for fats. Given these findings and the results of Drewnowski's experiments in the early 1980s, it is possible that we may find eating fats even more pleasurable than sweets.

It is known that humans and other animals can discriminate among different dietary fats and have a preference for corn oil, which can be used as a positive reinforcer in conditioning experiments. Corn oil has three major fatty acid components: linoleic acid (52%), oleic acid (31%), and palmitic acid (13%). Recent experiments in rats have shown that LA has an important effect on the physiological responses of the epithelial taste cells that make up taste buds. It appears that when LA binds to these cells, it increases the strength of the electrical signal that they normally send to the brain-stem in response to a food source. For instance, if LA and sucrose are consumed together, the combined signal sent from the taste bud that announces the arrival of food is stronger than would be the case with sucrose alone. This physiological response has a marked impact on food intake regulation. In a series of behavioral experiments, psychologist David Pittman and his students at Wofford College found that in rats, LA acts to increase the intensity of sweet, salty, and sour tastes such that the natural preference or avoidance of each is enhanced. As predicted from the physiological findings, animals preferred the taste of a solution containing LA and sucrose together more so than a solution with sucrose alone. Likewise, when Pittman's rats were given a mixture of LA with salt or citric acid, they consumed less than when the salt or citric acid solution was offered alone.

Linoleic acid is present in a variety of natural vegetable oils, and since it has a direct effect on the physiological responsiveness of epithelial cells, it is likely to be one of several compounds that give fats their pleasurable taste. The fact that LA can be used as a positive reinforcing stimulus in conditioning tasks tells us that humans and animals are motivated to consume foods that contain the substance. Hence, the pleasure we find in eating fats may serve to ensure that enough essential fatty acids are included in our typical diet to promote and maintain normal brain growth and development. At the same time, this pleasure-mediated mechanism provides yet another

example of how modern food manufacturing technology, in pro-
liferating the availability of refined sugars and fats, has essentially
removed the selection factors that originally led to these important
adaptations. In doing so, we are a society vulnerable to a number
of disorders, such as obesity and diabetes, that emerge when the
pleasure we receive from eating certain foods is filled well beyond
the natural limits imposed by the environmental circumstances of
younger hominids.

Chapter 7

The Evolution of the Lullaby

*Is it not strange that sheep's guts should hale souls
out of men's bodies?*
                    —William Shakespeare, *Much Ado about Nothing*

*As neither the enjoyment nor the capacity of
producing musical notes are faculties of the least use
to man in reference to daily habits of life, they must
be ranked among the most mysterious with which he
is endowed.*
                    —Charles Darwin, *The Descent of Man*

Several years ago I volunteered at a behavioral clinic and worked
with a group of fourteen adolescents who were diagnosed with
attention-deficit-hyperactivity disorder (ADHD). It was an amazing
thing to see such large differences in symptoms among teens with
the same diagnosis and even within the same individual from day to
day. We had our share of students who were chock full of what most
would consider normal childhood energy, along with others who

were clearly different in that their activity levels seemed unending in practically every context.

At our clinical staff meeting one Monday morning, a new intern raised the possibility of adding music therapy to our group sessions. While some of us, I'm sure, were picturing a guitar and perhaps a few harmonicas, she went on to tell us about her friend who teaches African drumming. She argued passionately that kids with ADHD respond well to drum sessions because they promote group cooperation and turn-taking. Within two weeks, we had fifteen beautiful instruments—ten *kpanlogos*, with their warm, earthy bass tones, and five *djembes*, offering a high, snappy timbre.

The drums came complete with a colorful music therapist fresh out of UC Berkeley, who visited us every Thursday afternoon. Joachin began a typical drum session by gathering all the students into a circle and starting a very simple rhythm consisting of one beat sounded roughly every second—a "heartbeat." In the first few weeks, just getting all fourteen students to sit in their chairs at the same time was a genuine accomplishment, but by the end of the first month they began to look forward to each session, and the changes in our little community were palpable. At some point during the fifth session, I remember feeling a deep sense of pride at how fast our circle had joined into a synchronized beat that particular day.

Joachin gradually introduced more complicated rhythms that were to be played on top of the heartbeat. Each student had a rhythm to maintain in concert with the entire circle—some had to play *menjani*, while others had to play *aconcon* or something else, and this varied from session to session. Above all, there was no room for solo performances in the circle, and whenever the group's overall rhythm broke down, we would start over amid sighs of frustration. Toward the end of the third month a noticeable improvement in our collective sound became obvious. The sessions began to take on a true communal feel, and at times the group's two or three main rhythms were so synchronized that the sound became almost hypnotic. During these periods I often lost myself in the moment, imagining the millions of other drum circles that have played across time and culture—humans of all kinds joining and celebrating nature's periodicity. One minute I'm drumming with students in a therapeutic residence, the next I'm part of a tribal clan of early hominids living along the Rift Valley. Perhaps we're drumming in preparation for a hunt or to celebrate the arrival of a newborn or a marriage.

Ancient drums have been discovered in almost every part of the world. Their earliest appearance in the archaeological record dates back to about 6000 B.C., excavated from Neolithic Era sites in northern Africa, the Middle East, and South America. Ceremonial drums have been found in these regions, along with wall markings depicting their use in various aspects of social and religious life. Other percussive and even flute-like instruments have been unearthed at *Homo sapiens* sites throughout Europe and Asia dating as far back as a hundred thousand years. And the music wasn't limited to modern humans—it appears that Neanderthals made music as well. Archaeologists excavating a cave near Idrija in northwestern Slovenia recently found a bear's polished thighbone with four artificial holes drilled into it that were aligned in a straight line on one side. Although we don't have any way of knowing if this sixty-thousand-year-old object was ever used to make sounds or even music, very similar "bone flutes" have been discovered at *Homo sapiens* sites and estimated to be forty thousand to eighty thousand years old.

The rule is very simple in modern cultures across the world and throughout all of recorded history: wherever there are humans, there is music. No recorded human culture—whether extinct or extant—has ever been without music production. Although what passed for a melody in ancient China undoubtedly differs from, say, what a twenty-first-century European might find entertaining, all humans have a faculty for producing and enjoying music. Indeed, given the omnipresence of music production and enjoyment across human civilizations, some researchers consider musicality to be an evolutionary adaptation, perhaps akin to language. But unlike language, which is used to communicate our thoughts to others, music has no clear-cut survival or reproductive consequences. So the question remains: What is the adaptive function of music?

There are generally three schools of thought on the origins of music. The first group views music as an interesting, albeit evolutionarily irrelevant artifact of our sophisticated brains—a form of "auditory cheesecake." The basic idea is that humans evolved a set of sophisticated cognitive, motor, and perceptual skills that have clear survival and/or reproductive value, and the expression of these skills led naturally to the emergence of other by-product abilities, such as art appreciation and musicality. The cheesecake view is overwhelmingly the most popular in mainstream psychological thinking today.

A second view is that music has real survival value and has been forged by the same principles of natural selection that have shaped other cognitive abilities such as binocularity, color vision, sound localization, and so forth. A wide range of suggestions for the function of music has been made, most having to do with its ability to bond the social group through coordinating action and ritual. Undoubtedly, music can have a profound influence on the behaviors and emotions of large groups of people—if you need to be convinced, simply visit a local nightclub or ballpark. This view, however, has its problems because it depends on the rather untenable position of invoking group selection to account for the evolution of musicality—a mechanism that has never proven convincing to scholars of mammalian evolution.

Finally, a third school of thought argues that music evolved primarily through mechanisms of sexual selection rather than natural selection. The chief difference between the two, of course, is that natural selection fosters adaptations that increase an organism's likelihood of survival, while sexual selection fosters adaptations that increase the likelihood of successful mating and reproduction. Both mechanisms impact the ultimate scorecard of evolution—how well an organism passes on its genes—yet the adaptations that emerge can often be at odds with one another. For instance, the size and color vibrancy of the peacock's tail is an important variable in reproductive success. Peahens are attracted to males with the largest and most colorful displays. At the same time, however, this conspicuousness puts "handsome" peacocks at a survival disadvantage from a natural selection viewpoint because they are easier to spot by predators, and vibrant, cumbersome tails make them less able to evade an attack. Indeed, adaptations driven by sexual selection often emerge *because* they handicap an organism's survival in some way that makes it easier to assess its true fitness. In this example, the fitness cost of having a large and colorful tail makes the peacock an easy target. Those males who have the most conspicuous tails are truly the fittest—the thinking goes—because they can afford both the metabolic cost of growing a large tail and the survival cost associated with attracting the attention of predators. In this context, music is seen as just another ornate animal display designed to get the attention of the opposite sex. The proponents of this view argue that music production is a reliable fitness indicator because it signals an ability to maintain a high degree of skill at the cost of diverting energy, attention, and time away from basic survival behaviors.

Each of these perspectives on the origins of human musicality is based on distinct mechanisms and therefore has unique implications for our relationship with music and why we find it pleasurable. In this chapter I will offer a fourth perspective: that our attraction to music results from a developmental requirement that we experience distinct classes of auditory stimulation for normal brain growth and maturation throughout life but particularly during the first two decades. We will find that there are innate constraints on musical sensitivity that transcend cultural differences and provide a core set of features common to all styles and genres. These features have a great deal in common with the singsong of motherese and will offer clues as to why we find pleasure in music as well as many other types of acoustic experiences.

A Universal Grammar

Music is said to be the universal language, but exactly what properties, if any, can be found that transcend culture, geography, and time? Most people prefer the musical genre they grew up with, and even the most casual observer must concede that there is tremendous variation in style from generation to generation. Clearly, learning plays a large role in shaping the specific musical idioms we prefer. Research throughout the past decade, however, has begun to show that certain sounds and note combinations have virtually universal effect on the emotions of listeners independent of the culture in which they were born, raised, and live. Moreover, most neurologically normal listeners, no matter where they are from, can agree on what is and is not musical, even when the sequence of tones is novel or drawn from a foreign scale. This has led some theorists to focus on the similarities between music and language development when speculating on the origins of musicality.

Decades ago, the linguist Noam Chomsky set out to understand why all normal children spontaneously speak and understand complex language. He pointed out that all mature speakers of a language can generate and interpret an infinite number of sentences, despite great variation in their levels of formal education. Moreover, in any given language, most native speakers can agree on whether a sentence seems grammatical. Since most speakers have these abilities despite varying levels of formal linguistic training, Chomsky argued that we are all born with an innate knowledge of language. The instinctual set of rules we unconsciously use to make grammatical judgments as well as to produce and interpret sentences is called the universal grammar. Chomsky argued that linguistic development involves the fine-tuning of this grammar toward settings appropriate to the indigenous language.

The composer Leonard Bernstein was the first to apply Chomsky's ideas about language to music. He suggested that all the world's musical idioms conform to a universal musical grammar. This theory was advanced more formally through the work of psychologist Ray Jackendoff and musicologist Fred Lerdahl. They viewed music as being built from a hierarchy of mental structures, all superimposed on the same sequence of notes and derived from a common set of rules. The discrete notes are the building blocks of a piece and differ in how stable they feel to a listener. Notes that are unstable induce a feeling of tension, while those that are stable create a sensation of finality or being settled. Musical styles differ in the emphasis placed on beat interval and pitch, but most genres use notes of fixed pitch.

Pitch is related to the frequency of the sound wave's vibration that is emitted by an instrument, but is perceived musically relative to other notes and the interval separating them rather than in any absolute sense. When a guitar string is plucked it vibrates at several frequencies at once: a dominant frequency called the fundamental and integer multiples known as harmonics, which add fullness and timbre. For example, a note with a vibration of 64 times per second will have overtones at 128 cycles per second, 192 cycles per second, 256 cycles per second, and so forth. The lowest frequency—which is often the loudest—determines the pitch we hear. In this example, the fundamental frequency is 64 cycles per second and corresponds to the second C below middle C.

When a sound wave vibrates faster, say at a fundamental frequency of 128 cycles per second, we perceive the tone as being higher. Since the fundamental frequency of this new tone at 128 cycles per second is related to the other tone at 64 cycles per second by an integer multiple ($128 = 64 \times 2$), it will sound higher but with the same pitch (a middle C). The interval that separates our two example tones at 64 and 128 cycles per second is called an octave. All primates perceive tones separated by an octave as having the same pitch quality. The pentatonic scale, common to most musical idioms across the globe, is built from having five distinct pitches within an octave. Throw in two additional pitches per octave and you have the seven-tone diatonic scale that forms the foundation of all Western music, from Beethoven to the Beatles.

Music is governed by a relatively small set of rules—like language— that can be used to generate an infinite variety of compositions. Music also employs recursion. In the same way that a sentence can be lengthened indefinitely by adding modifiers or additional words, so can a musical piece by inserting new or repeating phrasing. And just as language emerges naturally in children without a need for formal linguistic training, so too does music. Indeed, the only requirement for the development of musicality in babies is exposure to music.

As we have seen in earlier chapters, human newborns are far from being blank slates. With regard to the sensations of touch, motion, smell, and taste, they have clear preferences for certain stimulation patterns that are optimally tuned for regulating brain growth and development. The same is true for hearing. Newborns are attracted to music from birth and are sensitive to acoustic properties that are common to all music systems across cultures. By the time an infant is two months old, it will have roughly the same ability to distinguish pitch and timing differences in musical structure as that of listeners with decades of exposure to music. From the very beginning of life, newborns are attracted to specific features of music that are also preferred by adults the world over.

Babies as young as four months old show a stable preference for music containing consonant rather than dissonant intervals (an interval is a sequence of two tones). They also discriminate two melodies apart more easily if both have a consonant interval structure rather than a dissonant structure. Consonant intervals are those where the pitches (the fundamental frequency) of the constituent tones are related by small integer ratios. For example, intervals such as the "perfect fifth," with a pitch difference of seven semitones, or the "perfect fourth," with a pitch difference of five semitones, have very small integer ratios of 3:2 and 4:3, respectively. Adult listeners from all cultures find these intervals pleasant-sounding, and babies love them. Both adults and four-month-olds prefer these consonant intervals to dissonant intervals such as the tritone, with a pitch difference of six semitones and a large pitch ratio of 45:32. Infants listen contentedly to melodies composed of consonant intervals but show signs of distress when some of the intervals are replaced by dissonant intervals. This effect has been observed in many cultures and in infants with varying levels of music exposure. Hence it appears to result from an innate predisposition toward certain acoustic features that are pleasurable and indeed seem to be shared by most systems of music.

Another interesting feature of auditory processing that is present in infants is their ability to detect transpositions of diatonic melodies across pitch and tempo. Both infants and adults can recognize a tune based on the diatonic scale as the same when it is transposed across pitch, but fail to do so when the melody comes from a nondiatonic scale. Since all primates perceive tones separated by an equal octave as having the same pitch quality, one might predict that the ability to detect transposition of diatonic melodies is also present in our hairier cousins. To date, this experiment has only been performed with rhesus monkeys and, as expected, they exhibit the same effect as human adults and infants. The presence of similar auditory preferences and perceptual abilities among adult listeners and infants from different cultures suggests that certain features that are critical components of music competence exist at birth.

Eenie-Meenie-Miney-Mo

That certain auditory biases exist at birth is probably not news to parents. Even those who are uninitiated to this phenomenon learn quickly that their prelinguistic newborn is a capable communicator. Infants communicate with emotional expression, and parents use this to gauge what their child needs. Few stimuli calm an infant and get their attention more effectively than the lullaby sung in a soothing voice. As we saw in earlier chapters, infants recognize their mother's voice from birth, and are calmed when they hear it. Experiments have shown that newborns and infants are highly sensitive to the prosodic cues of speech, which tend to convey the emotional tone of the message. Prosody is exaggerated even more so in the typical singsong style of motherese that dominates parent-infant dialogue during the first year of life. The infant trains its parents in motherese by responding positively to certain acoustic features they provide over others. Motherese and lullabies have so many acoustic properties in common—such as simple pitch contours, broad pitch range, and syllable repetition—that theorists have argued them to be of the same music genre.

Just as motherese shows up with the same acoustic properties in virtually every culture, so too does the lullaby. Practically everyone agrees on what is and is not a lullaby. Naive listeners can distinguish foreign lullabies from nonlullabies that stem from the same culture of origin and use the same tempo. Of course, infants make the distinction quite readily. Even neonates prefer the lullaby rendition of a song to the nonlullaby rendition performed by the same singer. Although it is tempting to attribute such preferences to experience, studies have shown that hearing infants raised by deaf parents who communicate only by sign language show comparable biases. It appears, then, that from our very first breath, we carry a set of inborn predispositions that make us seek out specific auditory stimuli. These stimuli are common across cultures and appear in many forms of music but are exemplified in the lullaby. Why should this be the case? One might argue that these acoustic features help foster mother-infant communication, but this just passes the question along without really answering it. Why do these specific acoustic properties show up in motherese and the lullaby? They arise because the infant trains his or her parents to provide these stimuli through feedback in the form of emotional expressions of approval and calm. The real question is why these types of auditory experiences pacify and bring pleasure to infants (and adults).

In April 2003, scientists from the University of California at San Francisco discovered that newborn rats fail to develop a normal auditory cortex when reared in an environment that consists of continuous white noise. The hallmark of white noise is that it has no structured sound—every sound wave frequency is represented equally. Neurobiologist Michael Merzenich and his student Edward Chang wanted to understand how the environmental noise that we experience every day influences the development of hearing disorders in children. They speculated that perhaps the increase in noise in urban centers over the past several decades might be responsible for the concomitant increase in language impairment and auditory developmental disorders observed in children over the same period.

Their experiment began by raising rats in an environment of continuous white noise that was loud enough to mask any other sound sources, but not loud enough to produce any peripheral damage to the rats' ears or auditory nerves. After several months, the scientists tested how well the auditory cortex of the rats responded to a variety of sounds. They found significant structural and physiological abnormalities in the auditory cortex of the noise-reared rats when compared to rats raised in a normal acoustic environment. Interestingly, the abnormalities persisted long after the experiment ended, but when the noise-reared rats were later exposed to repetitious and highly structured sounds—such as music—their auditory cortex rewired and they regained most of the anatomical and physiological markers that were observed in normal rats.

This finding created a wave of excitement throughout the scientific community because it clearly showed the importance of experience in influencing normal brain development. The developing auditory cortex of all mammals is an experience-expectant organ, requiring specific acoustic experiences to ensure that it is wired properly. As Chang summarized, "It's like the brain is waiting for some clearly patterned sounds in order to continue its development. And when it finally gets them, it is heavily influenced by them, even when the animal is physically older."

The auditory cortex of rats and humans—indeed, all mammals—progresses through a very specific set of timed developmental changes. As we have seen in the other sensory systems, this development depends on genes to program the overall structure, but requires the organism to experience environmentally relevant stimuli at specific times to fine-tune the system and trigger the continued developmental progression. Genes don't just magically turn on. In most cases they wait for an internal or environmental promoter to trigger their expression. And the details of development are not in the genes but rather in the patterns of gene expression.

The primate auditory system develops a bit differently from the sensory systems of touch, smell, and taste that we have considered thus far. The peripheral anatomical structures of the auditory system begin to form very early in development, yet the system matures rather slowly as a whole. For instance, by the time Kai had been in Melissa's womb for about four weeks, he already had the beginnings of ears on either side of his embryonic head. Cells were also forming in what will become his cochlea, the shell-shaped organs in each ear that transduce acoustic sound waves into electrical impulses that the brain uses to communicate. By about the twenty-fifth week of gestation, Kai had most auditory brain-stem nuclei in place that will be used to process features of acoustic information such as sound localization and pitch discrimination. But these cells depend on stimulation for continued growth, maturation, and being able to form synaptic connections with their higher cortical target sites.

It is probably no surprise to readers by now that it is precisely at this time—when the brain most needs auditory stimulation—that fetuses begin to hear their first sounds. We know this for two reasons. First, it is at this age that fetuses first show signs of what is called an auditory-evoked potential. Preterm babies are given a battery of tests. Among these is a painless test that involves placing a small headphone over their ears and attaching three electrodes to their scalp to measure their brain's response to auditory stimuli. When a brief clicking noise is played, preterm babies younger than about twenty-seven weeks show little or no electrical response following the stimulus—their brain is not mature enough to register the sound, and they show no sign of hearing it. It's not until after twenty-seven weeks or so that preterm infants show the first evidence of a brain response to auditory stimuli, and not so coincidentally, the first signs of actually hearing sounds.

These results are consistent with observations using ultrasound technology to monitor fetal movements in response to tones played on their mother's stomach. At Kai's sixteen-week ultrasound, he showed no response to auditory stimulation in the form of tones played near Melissa's stomach, or either of our voices. The story had changed by his thirty-week ultrasound. Not only did he appear less embryonic, he also altered his movements whenever we made a loud noise. The most reliable change was a complete halt of his ongoing movement when his mother spoke. My paternal observations are consistent with real experiments showing that fetuses start and stop moving in response to auditory stimuli, and even blink their eyes in reaction to loud sounds heard in the womb.

Throughout the last trimester, Kai's brain was taking in sounds and using them to stabilize and fine-tune his developing auditory system. Although many sounds can pass through to the womb, he was especially sensitive to those that changed with dramatic pitch contours. This is because even fetal brains show adaptation to unchanging stimuli. A tone that is repeatedly played at the same pitch and amplification is responded to fully at first, but becomes less and less interesting over time. This is mirrored by physiological responses measured from the brain such as auditory-evoked potentials. Evoked potentials become smaller and smaller in preterm babies if the same old boring stimulus is played over and over again. The brain simply begins to habituate, and the stimulus becomes less salient.

Continuous and slowly changing sounds—those that exhibit exaggerated pitch contour and wide pitch variation (exactly like those heard in motherese and in lullabies)—keep the baby and its brain in an attendant state. Fetuses show far less behavioral habituation to music that sounds like motherese than to repetitive tones of the same exact pitch. Likewise, preterm infants older than thirty weeks do not exhibit a decline in their auditory-evoked potential if they are stimulated with sounds that change slightly in pitch rather than stay the same. The sounds of motherese and lullabies are born from acoustic features that are the perfect forms of stimulation to ensure that a fetus's experience-expectant brain will continue to develop normal auditory circuitry and perceptual skills that will help it survive after birth.

The auditory system is not the only part of the brain that benefits from sound stimulation. Research has shown that fetuses older than thirty weeks can distinguish different phonemes such as *ba* versus *bi*, suggesting that prenatal experience may be critical to the development of language areas. There is also evidence that auditory stimulation while still in the womb promotes the development of limbic structures such as the hippocampus and the amygdala that support memory and emotional development. Indeed, it is now clear that the sounds a fetus hears in its third trimester can be remembered years later and even influence behavior as late as two years after birth. One researcher, for example, found that infants whose mothers watched a particular soap opera during pregnancy were calmed when they heard the show's theme song, whereas babies whose mothers did not watch the show had no reaction to the song.

Now that Kai is finally born, he is awash in a sea of acoustic information, but not all of these sounds are novel. He is certainly familiar with his mother's voice and to a lesser extent my own. Many of the sounds that Melissa experienced in her final trimester were likely heard by Kai, and although most were not repeated enough to consolidate into long-term memories, they undoubtedly had a significant impact on his auditory development thus far. Kai, like all primates, will continue to need auditory stimulation for decades to come. Normal development of auditory circuitry continues well into the late teens, resulting in steady improvement in many functions such as pitch discrimination and sound localization. Mammals that are denied this stimulation suffer from a range of abnormalities. For example, rats that are raised in an acoustic environment with a restricted frequency range are unable to hear outside this range as adults. This impacts their ability to discriminate sounds that have pitch variation that overlaps with this frequency range. Deprivation also disrupts their ability to localize sounds—an impairment that could prove costly if approached by a predator.

The fact that all primates have auditory perceptual skills that are facilitated by diatonic scale structure, while not true for all mammals, gives us a rough idea of when our faculty for music may have emerged in our evolutionary lineage. Some Old World primates may have evolved auditory circuitry that had improved function relative to competing primates—such as increased pitch discrimination and sound localization—that gave them a distinct survival advantage.

As we've seen with modern experimental studies, the successful development of this circuitry depended on the organism experiencing certain forms of auditory stimulation. Clearly, not all primate species have satisfied this demand in the same way. In hominids, natural selection has forged this adaptation by linking these optimal forms of auditory stimulation to the activation of evolutionarily ancient pleasure circuits that are seen in all mammals. These circuits were most likely an earlier adaptation that fostered reproduction. Natural selection produces incremental change in structure and function that is always built on top of earlier adaptations. Structures are co-opted from others not in a design sense, but through a process that unevenly results in the survival of some genes over others. Hominids' new fondness for wide swings in pitch variation and loudness in combination with exaggerated intonation may have created the initial conditions that ultimately led to the evolution of musicality and motherese in our species. These human technologies, in turn, became very effective tools in promoting brain development. In the next chapter, we will find that a similar story has occurred for vision.

Chapter 8

In Search of Pretty Things

One eye sees, the other feels.

—Paul Klee

The real voyage of discovery consists of not in seeking new landscapes but in having new eyes.

—Marcel Proust

Martin watched the first flurries of snow begin to fall and finally relaxed as he switched off the outside light and urged his body upstairs to bed. The workday had been a frenzy of meetings and deadlines blurred into a whirl that began when he entered the freeway on-ramp in the morning and subsided only after a glass of wine before dinner. His drive home had been the pièce de résistance of the day. Another car seemed to emerge from thin air as he was changing lanes, causing him to veer suddenly. The remainder of his commute was accompanied by white knuckles and a sickening cold sweat—the holidays always seemed to bring out the amateurs.

No matter how tense his day, Martin never seemed to have any difficulty getting to sleep once his head hit the pillow. His last thoughts on this cold December evening were of his beautiful wife and his four-year-old son's warm, crooked smile—a sight that always overwhelmed him.

Well before daybreak Martin was roused from sleep for his customary middle-of-the-night trip to the bathroom and immediately felt the gripping headache. Stumbling into the bathroom, he noticed a strange numbness in his left leg and the left side of his face. In the dim glow of the bathroom night-light, he saw that something was also wrong with his face. It was hard to say exactly what was different at first, but slowly he realized that his mouth seemed at odds with the rest of his face—the left side kind of drooping placidly. Worse yet, when he moved his right arm toward the drug cabinet, his hand seemed to vanish from sight, only to reappear once it came in contact with the mirror. Later that morning in the hospital, his doctors had a hard time convincing him that although just forty-one years old, he had experienced a stroke that could have ended his life.

After six months of physical therapy, Martin began to feel like himself again. The stroke had been a warning, but he'd survived and come through fairly unscathed save for occasional slurred speech. The new Martin had been molded into a different person altogether—finding more time for relaxing with his family, and even for a daily jog around his neighborhood each morning before work. All of his attending doctors pronounced him fit—and lucky—yet he often felt odd sensations during the day. More and more frequently during his morning jogs, Martin seemed to have brief moments of intense fear and panic. These fleeting episodes occurred especially when he ran through the park at the end of the street and were often triggered by the sudden appearance of a dog—any dog, no matter how small—or even another person walking toward him.

Although by all standard tests Martin had excellent vision, he mentioned during a visit with his doctor that he occasionally had trouble estimating the distance of approaching cars while crossing a street—a problem that had led to several close calls in the past few months. When asked about any other difficulties, he reluctantly confessed that

sometimes when talking with a coworker he had difficulty understanding what she was saying, because her mouth seemed to "fade in
and out" of sight when she spoke. This must have been a surprise to
the attending neurologist, but it was enough of a clue to raise the possibility that his patient was suffering from *akinetopsia*, or visual motion
blindness.

Martin was referred to a specialist that week for a battery of
visual acuity, motion, and perimetry tests. His static visual acuity was
completely normal. In tests using moving stimuli, however, Martin
seemed to have severe deficits in specific portions of his visual
field. Objects at rest that began moving proved to be particularly
troublesome.

A common test for akinetopsia involves what psychologists and
other vision scientists refer to as object tracking. The patient is seated
directly in front of a computer display and asked to focus his or her
gaze and follow the movement of a colored circle. The circle begins
in the middle of the screen, and the tester can manipulate the object's
gradual movement to different corners of the screen. People with
normal vision see the circle move smoothly from the center to the
corner locations, and grow bored quickly. Patients with akinetopsia
have an entirely different experience. They report seeing the still
object disappear from the center and reappear in one of the corners.
Once the circle moves, it is as good as gone until it comes to rest.
Martin had a similar experience in his object tracking test, so his
neurologist immediately ordered a second MRI scan of his injured
brain. The damage was concentrated on one side of Martin's posterior parietal lobe, an area toward the back top of the brain that is
deeply involved in—not surprisingly—motion perception.

The fact that Martin had stroke damage to his posterior parietal
lobe provides an explanation for his akinetopsia, yet the particular
way this disorder impacted his life—creating a sudden, blinding fear
of dogs and people coming toward him—can only be understood
if we consider how the pleasure instinct and experience guide the
development of the visual brain.

How Pleasure Fine-Tunes the Visual Brain

The emergence of the earliest primates from the mammalian branch some sixty million years ago came with dramatic changes in the sensory systems of this lineage. Based on the abundance of fossils these animals left behind, we know that they were rather small—probably weighing only a few ounces—and closely resembled modern-day prosimians such as galagos, tarsiers, and lemurs. They were adapted to a much warmer climate than we have today, with tropical rain forests covering a significant portion of the planet. Their small size and prehensile hands and feet allowed them to grab onto and forage among the fine terminal branches that make up the rain forest canopy. This unique niche came with its own challenges in terms of identifying potential foods, usually fruit, seeds, and insects camouflaged against a background of green leaves and thickets, and for recognizing potential predators. These two key selection factors—needing to locate hidden fruit and predators—promoted the gradual shift in a sensory system dominated by smell in most mammals to a new model where vision reigned supreme in the emerging early primates. These new creatures had large, forward-facing eyes with a high density of photoreceptors in the center of their retina, an area called the fovea. In early primates this high concentration of photoreceptors came with new brain-stem circuitry that evolved to focus their visual gaze frontally toward anything that moved, and a dramatic increase in the size of the brain areas devoted to vision relative to those devoted to olfaction.

With the shift toward frontal vision, early primates sacrificed some ability to detect the presence of food or predators using smell, yet these anatomical changes gave them distinct advantages over other groups of mammals. In particular, the shift from side-oriented eyes to a frontal position permitted the evolution of binocularity and stereoscopic vision, two functions critically important for fine visual acuity and determining the size and distance of an object.

The favoring of vision over smell in early primates was not just a matter of the visual cortex getting bigger—entirely new brain areas devoted to specialized visual functions evolved in these animals that never existed in other mammals. One important innovation was the evolution of brain areas in the posterior parietal lobe and medial temporal regions that were devoted to the visual guidance of muscle movement. Evolutionary biologists have argued that the presence of improved frontal visual acuity and a propensity to live among fine tree branches necessitated the development of neural systems designed to improve eye-hand coordination. The emergence of posterior parietal areas for perceiving visual motion is the result of this evolutionary ratchet effect (chapter 2), where one adapted function served as a selection factor for yet another adaptation. In this case, the development of increased frontal visual acuity in combination with prehensile hands and feet made the rain-forest canopy a viable niche for early primates. This shift from ground-level to tree-branch foraging incurred a high survival cost on primates with poor eye-hand coordination. Simply moving from one swaying branch to another might prove fatal if the distance to the next limb or its degree of movement was miscalculated. Such conditions served as strong selection factors in driving the evolution of specialized brain regions devoted to perceiving object motion.

The evolution of brain regions in the posterior parietal lobe and medial temporal areas that facilitate object tracking and eye-hand coordination, in turn, created the conditions in which selection factors arose for the development of color vision. It is thought that until about forty million years ago early primates had only one primary photoreceptor type tuned to a single distribution of light wavelengths. In terms of function, this mechanism allowed a primate to see the world in basic shades of gray. Most modern-day prosimians have two distinct types of photoreceptor (dichromatic) that are maximally sensitive to different light wavelengths. These animals have a rudimentary capacity for color vision. During the early period in the evolutionary history of primates, a mutation caused the genes that normally control photoreceptor development to duplicate. This process resulted in a higher density of photoreceptors, which eventually diverged into two and then three distinct classes, each sensitive to a different wavelength of visible light. Recent work has shown that

the advance from dichromacy to trichromatic color vision specifically enhanced the ability of primates to distinguish nutrient-rich
fruits from the background coloration of leaves. Thus the evolution
of improved eye-hand coordination and frontal vision that gave early
primates a novel niche created the conditions where identifying colored fruits against a uniformly green background of leaves served as
a selection pressure for color vision.

The evolutionary history of primate vision is a story riddled with
ratchet effects, and so too is the ontogenetic development of vision
in humans. By the fifth week of gestation, human embryos show the
first signs of an early eyecup that begins to differentiate into a lens
and a retina. At this point in development, the eyes face laterally to
each side, much like an early mammal. The retina itself is derived
from the same neural ectoderm that comprises the central nervous
system and is therefore considered part of the brain. Each retina is
attached to the brain-stem by a broad group of fibers called the optic
nerve. By the fourteenth week of gestation, the eyes begin to face
forward in the familiar primate form, and photoreceptor cells start
to form in the center of the retina and gradually fill in from the
fovea outward. Development continues from the retina inward to
brain areas, advancing along the same path as normal sensory input.
Retinal development is followed by the emergence of several brain-
stem sites such as the superior colliculus (involved in controlling eye
movements), then the visual cell groups in the thalamus (e.g., lateral
geniculate nucleus), and finally neocortical areas such as the primary
visual cortex (also known as V1).

Primates have tremendous developmental investment in vision.
There are more than forty known cortical areas devoted specifically
to visual information processing. By the end of the second trimester,
a human fetus has a fairly mature primary visual cortex, and most
secondary and tertiary visual centers have undergone rapid growth
both in terms of cell numbers and synaptic connections between the
areas. A human's visual system, however, does most of its development
after birth and is even more dependent on experience for normal
maturity than the other senses. While the other sensory systems can

be stimulated fairly early in the womb and thus undergo considerable experience-expectant growth before birth, vision is the exception to the rule. A scarcity of light makes its way to the fetus, and consequently most of the visual system's fine-tuning must be guided by the pleasure instinct after birth.

Any parent will tell you that infants are hungry for visual experiences, yet they are rather particular in their choices. Much like the other sensory systems, there seems to be a characteristic sequence of preferred stimulation types that is attractive for infants. Little Kai, who is now crawling, has followed a pattern of visual development that is an echo of primate phylogeny. At birth his visual system was relatively immature when compared to the sensory systems responsible for touch, taste, smell, and hearing. For the first few weeks, Kai could only lock onto faces and high-contrast objects that were within about five to ten inches of his face. Infants this age have great difficulty focusing on objects outside this range, since the muscles that control lens shape (which affects light refraction) are so immature. By four months, however, Kai could focus his vision across a much larger range of distances and was fascinated by anything that moved. He was particularly fond of ceiling fans, but the love affair always ended once it was turned off, suggesting that it was the motion that piqued his interest. Toy manufacturers are, of course, sensitive to these developmental milestones. One might argue tongue in cheek that the preference of newborns for things that move was an important (albeit artificial) selection factor in the evolution of the mobile (insofar as parents tend to select toys that are attractive to newborns).

Similar to what we have seen with the other sensory systems, the neocortical areas devoted to vision tend to specialize. Once information enters the first neocortical area dedicated to vision (primary visual cortex or V1), it diverges to a number of additional cortical regions, each with a different specialty. Some areas are responsible for processing object motion and location. Others process object form, color, texture, shading, shape, and so forth. Parallel processing is a general operating principle for all sensory systems in the brain. The strength of parallel processing lies in the fact that it facilitates speed of information transfer through the brain and adds to the robustness of the information through built-in redundancy (that is, multiple channels are used).

As I throw a multicolored ball in the air in front of my son, the information makes its way through Kai's retinas and quickly stimulates brain-stem sites, such as the superior colliculus, that focus his attention toward the object. This brain-stem attention-grabbing circuitry is about as evolutionarily ancient as any primate brain region, since it is found in every vertebrate. From the brain-stem sites, the visual information about the ball travels through Kai's thalamus and enters his primary visual cortex, which provides him with the first conscious perception of the object. Thus, although the brain-stem activation produced a change in behavior causing Kai to focus on the ball, this processing is beneath the surface of consciousness.

Once the visual information reaches V1, it quickly diverges into two dominant streams: one responsible for processing information about an object's spatial location and movement (the "where" pathway) and another responsible for processing the form features of the object such as shape, color, and texture. The latter stream is known as the "what" pathway.

The "where" and "what" visual pathways develop and mature at different rates in humans. In primates the brain pathways devoted to processing object motion develop and mature far earlier than those responsible for processing advanced object form information. For example, while Kai was delighted by the appearance of almost any slowly moving object when he was four months old, he is now a ten-month-old clearly in love with bright, primary colors. This is not really a transition from preferring objects that move to preferring objects that have bright colors. Rather, it is the addition of a fondness for bright, primary colors that joins the list of pleasure-inducing forms of stimulation. This sequence is characteristic of all human infants and maps onto the relative maturity of the "where" and "what" pathways.

At four months an infant's posterior parietal areas that comprise the where pathway are just beginning to undergo a major increase in synaptic pruning. Remember from earlier chapters that during synaptic pruning, experience becomes the crucial instrument in shaping and fine-tuning brain circuitry in the early stages of development. Hence a four-month-old infant's where pathway is just beginning to enter a phase of synaptic pruning, where its continued development and fine-tuning depend on appropriate stimulation. In this case, appropriate stimulation consists of any experiences that would optimally activate the mature circuit—namely, moving objects.

If we were to design this process in a robot—creating a sensory perception system that depends on experience for fine-tuning—any good engineer would build in a process to increase the probability that the optimal forms of required stimulation are experienced at the right times. Likewise, nature doesn't rely on the mere dumb luck that a developing infant will just happen to encounter specific forms of stimulation that are required for normal development. Nature has solved this problem by linking the brain circuitry that supports natural reward (primary reinforcing stimuli) with the maturing circuitry from the primary sensory systems. For example, to make the experience of motion perception pleasurable to Kai at four months, the growing circuitry in his posterior parietal lobe begins to establish reciprocal connections to several brain-stem and limbic regions that are involved in natural reward, motivation, and analgesia. Thus the activation of Kai's posterior parietal lobe circuitry by a slowly moving object such as a ceiling fan begins to be accompanied by pleasurable sensations much like those corresponding to primary reinforcing stimuli (such as sweets). This process ensures that Kai naturally seeks out objects that fill this experience-expectant requirement for the successful fine-tuning of his visual where pathway and the continued refinement of his capacity to discern motion.

Our crawling ten-month-old Kai is now entering a phase where some areas in his what pathway are undergoing extreme synaptic pruning. At about this time regions such as V8, devoted to processing color vision, begin to mature and consequently need proper stimulation for continued growth and refinement. Kai's emerging attraction to objects that are composed of bright primary colors—reds, greens, and blues—will encourage him to seek out these optimal forms of stimulation that provide the fine-tuning needed in region V8. Indeed, the primary colors correspond to the wavelengths of light that optimally activate distinct classes of brain cells in region V8. If these cells are damaged in an adult, for instance by a stroke or related trauma, the result is a complete loss of color vision with no change in other features associated with visual acuity. The developmental pattern that is seen in infants—object motion pathways maturing before most brain areas involved in visual object recognition—echoes the evolution of vision in primates. Comparative studies suggest that the object motion pathway evolved well before most brain regions that are devoted to object recognition. For instance, the object motion pathway is observed in all mammals, but features that support object recognition such as trichromacy did not appear until the divergence of the primate lineage from other mammals.

The Pleasure of Learning

Vision is no different in terms of general developmental properties than any other sensory system. Genes play a direct role in mapping out the major brain regions dedicated to vision and the general pathways that connect them. Somewhere in the mere twenty-five thousand or so genes that comprise the human genome, there is enough information to ensure that the enormously complicated wiring of the human brain (and the rest of the body, for that matter) is mapped out. Genes do not code specific paths, but rather cause the development of unique molecular markers that are used by growing brain fibers as targets. This process gets the connections approximately right but leaves the remainder of the job—the fine-tuning—to experience.

Fine-tuning the visual system takes a long time. While the neural pathways that mediate motion processing mature fairly early, the circuitry responsible for higher-order processing and detailed visual acuity continues to be fine-tuned by experience well into the toddler period. Like the other sensory systems, brain cells that comprise visual circuitry are not mini-blank slates waiting to be written upon. They come preprogrammed with certain receiver biases from the very beginning. Experiments done in the early 1960s showed that neurons in the primary visual cortex, visual thalamus, and even in the retina itself respond optimally to certain forms of stimulation and are barely activated at all by others. For instance, many cells in V1 tend to respond to straight lines of a particular orientation. If we could record from cells in your primary visual cortex right now, we could perform the following experiment. Imagine I display a completely white screen directly in front of your eyes. While you focus on the screen, I lower a pencil held by its tip until it enters your visual field. Light reflected off the pencil enters your retina, where photoreceptors transduce the light energy into electrical impulses that are then sent along the visual pathways we have been discussing. At each stage in the processing of this image, some cells respond to this particular form of stimulus; however, most remain quiet. Starting with your retina, then your thalamus, and including V1, only a select group of cells have a preferred tuning for this specific orientation of the pencil. Other V1 cells are sensitive to straight lines, but they will only respond when the pencil is rotated to their preferred orientation (for example, horizontal instead of vertical).

Interestingly, our V1 cells are not ordered haphazardly, but have a strict anatomical organization that is related to the degree of angular rotation of a viewed edge. All of the cells can be activated by a straight line, but the line has to have the correct orientation to excite a given cell and get it communicating with other neurons. The functional consequence of this physiological arrangement is edge detection, a capacity that is critical for many aspects of vision, such as identifying the natural boundaries of an object. Many brain theorists envision a scenario where information from multiple edge detector cells is integrated at higher cortical areas to form a representation of the entire object. Experimental evidence shows that multiple cells from V1 with different orientation tuning converge on the same neurons at higher cortical areas such as V2, so this theoretical position has anatomical support.

Edge detection by V1 cells tuned to straight lines of a particular orientation is just one example of preexisting biases that are built into brain cells from the earliest period of development. Tuning biases like this have been found in newborn brain cells in virtually every mammalian species tested. Interestingly, although most cells in V1 have a signal preference shortly after birth, experience plays a critical role in shaping the tuning to match the particular ecological niche that an organism inhabits. In most species, a critical period exists in the earliest stages of visual development where if V1 cells are denied stimuli that match their preferred orientation, they may die or be retuned to another orientation. If the cells are stimulated by the appropriate signals during this period, however, the tuning of the cell becomes increasingly sharpened and specific to the original bias. This has two important effects. First, the increased tuning makes information transfer less noisy, since variation in terms of what kinds of stimuli may excite a cell naturally decreases. A second consequence of this process is that while some signals may be detected quite easily with minimal stimulation, other signals that are slightly different from the preferred tuning will be completely missed. Hence, existing biases that are in place at or near birth can become magnified with experience, while others may die out or even be replaced. Individual brain cells are far from being blank slates at birth.

The critical periods for visual system tuning—like the other sensory systems we have encountered—occur when that particular circuit is undergoing synaptic pruning (see chapter 3). During this period brain cells increase their sensitivity to some forms of stimulation and necessarily lose their responsiveness to others. The sharpened tuning of cortical cells that support visual perception results in increased visual acuity for some features and a decrease for others.

Experiments since the 1960s have demonstrated that cats and monkeys who are denied visual stimulation in a particular eye during this period of extreme plasticity have marked visual deficits as adults. Moreover, the primary visual cortex (and other visual areas) of the deprived animals looks very different from that of normally reared controls. Usually there is an approximately equal portion of visual cortical area in V1 devoted to processing information from each eye. If one eye is covered during the critical period, the portions of V1 that receive information from the competing eye expand and take over the areas that would have been associated with the covered eye. As we have seen in earlier chapters, synaptic pruning is a process that is ruled by competition. Two synaptic connections vying for the same space will each struggle to stabilize into a mature circuit, but the one that is activated by visual experiences the most usually wins. The old adage "Use it or lose it" rings true in this case. With the appropriate stimulation (nurture), the initial visual circuitry laid down by nature is further shaped to match environmental contingencies.

When cats or monkeys are reared in a carefully controlled environment where they only experience lines of a single orientation (for example, all vertical or all horizontal) during the critical period, their V1 cells stop responding to other orientations and retune to fire only at the experienced orientation. Later, as adults, these animals show poor visual acuity for detecting edges at novel orientations relative to control animals.

Humans also show signs of this effect. For instance, at least one study has demonstrated that North American Indians reared in traditional teepee-shaped dwellings have better visual acuity for oblique or diagonal angles when compared to people raised in "carpentered" environments (that is, house and apartments) that are filled predominantly with vertical and horizontal orientations. We begin with a set of preexisting preferences for visual scenes, but there is considerable latitude in how early exposure can retune brain cells that support vision and the connections between them.

Neural Bootstrapping

At birth, humans have a visual acuity of about 20/600, which is roughly thirty times poorer than that of normal adults. The attraction babies have for high-contrast objects and faces provides just enough stimulation for growing visual cortex cells to continue to mature at a reasonable pace. In the first three months, visual acuity steadily increases. Infants become more attracted to even finer gradations of contrast and are particularly fond of contrasting patterns that have pronounced lateral symmetry. It is not until their ability to experience these more nuanced visual patterns occurs that a second period of tremendous growth and synaptic pruning kicks into gear in V1 and higher cortical areas that are responsible for so-called hyperacuity.

Theoretical calculations of visual acuity based on the actual physical size and density of photoreceptors suggest that we should not be able to see as well as we do. Higher cortical areas, however, have a rich bag of tricks for solving visual problems like completing object patterns from partial or obscured inputs. These mechanisms radically improve visual acuity beyond the expected theoretical limits.

Interestingly, the development of hyperacuity depends more on experience in the second six months than the first. This is because humans must first experience simple visual patterns that promote the development of subcortical areas and V1. Once this circuitry matures in the first six months, babies become increasingly attracted to richer patterns of visual stimulation, such as scenes with more subtle contrasts and strong lateral symmetry. These experiences are, in turn, needed for the stimulation and normal maturation of higher cortical regions during their growth spurt in the second six months that support hyperacuity. This developmental pattern is so lawful that pediatricians often use visual tests of hyperacuity as an indicator of normal brain growth at twelve months.

This sequence is yet another example of how the brain bootstraps its own development. When workers construct a suspension bridge, they first extend a thin cable across the body of water. They then use that small cable to hoist a larger one, followed by a third, and so forth. Before long they have created a thick cable of intertwined wires that can support a Friday afternoon rush hour. The brain does something quite similar in its development. Just enough physiological maturation occurs to facilitate functional capabilities that, in turn, permit stimulation for the next phase of development. We see this modus operandi again and again in brain development—across every sensory system.

Bootstrapping as a mechanism for individual development is somewhat analogous to ratchet effects in phylogenetic development. In an earlier example, we saw how the evolution of visual motion processing paved the way for early primates' ability to forage among the fine branches that comprise the rain-forest canopy. Living in this new ecological niche created selection pressures for being able to clearly identify pigmented fruits and predators against a background of green forest. Compelling evidence from comparative studies demonstrates that trichomacy evolved in primates as a response to such selection pressures.

Bootstrapping is a highly efficient way to drive development, given that only twenty-five thousand or so genes are available to code the staggering amount of information required to grow a human. Clearly, not every step in development is written into the genes. Bootstrapping as a general developmental mechanism requires only information about the starting conditions and a means to encourage organisms to seek out the appropriate forms of experience necessary to stimulate further growth into the next stage.

A legacy of this process is that adult humans are strangely drawn to the same forms of visual stimulation that supported their brain development as newborns, babies, and toddlers. I'm not implying that the average adult finds pleasure in spending hour upon hour watching a ceiling fan or Big Bird look for Ernie. We are, however, attracted to the same general patterns of stimuli that bootstrapping *required* for normal brain development. This, of course, is manifested in a diverse number of ways in adults, some of which are undoubtedly flavored by cultural conventions.

Humans across different cultures are attracted to bright primary colors, scenes with pronounced lateral symmetry, and high-contrast objects. These biological preferences have an important influence in shaping what we find attractive. The advertising industry has been aware of these biases in our sensory processing for decades and designs product packaging that taps into these preferences. Often we are not even consciously aware of why we are attracted to a product. Keep in mind, however, that conscious awareness is not something evolution cares about. My son Kai does not need to *know* that he is attracted to faces and objects with bright colors for these forms of stimulation to benefit his visual development.

In the parlance of evolutionary biology, such innate preferences are sometimes called *receiver biases*. The term comes from an analogy in signal theory and has been applied to studies of animal communication. Modern communication devices can be built in two very different ways. One method, for example, is to develop a general device that uses a wide range of electromagnetic frequencies—much like the AM/FM radio. A problem with this general approach, however, is that the sender and receiver both have to be on the same frequency for communication to occur. There is a certain amount of luck in this happening, since the probability of the receiver and sender naturally sharing a channel decreases with an increasing number of frequencies.

An alternative approach is to build a receiver that is pretuned to specific frequencies. In this scenario, a radio will only pick up one or two selected frequencies, but with little potential for interference from other signals. Broadcasters who want to reach listeners with these radios will, of course, have to use devices that are specifically tuned to send signals at these frequencies. Only broadcasters who can send signals at the selected frequencies and with sufficient intensity will be successful in their communication attempts.

There are abundant examples where nature has followed suit—adopting either of the two strategies in intraspecies communication. The second approach is intriguing for our present discussion because receiver biases can arise from any number of sources. There are compelling examples where a physical feature that is found to be attractive by one sex and the preference for that feature by the opposite sex did not coevolve via genetic correlations. That is, a mating preference for a feature can sometimes emerge from developmental constraints rather than adaptations related to reproductive success.

It's important to point out, however, that most receiver biases are probably not associated with pleasure. For instance, Bolivian anuran frogs have an auditory system that is tuned to best hear vocalizations such as mating calls at 800 hertz. This particular receiver bias is the result of the physical properties of mature hair cells embedded in a frog's cochlea. In this case, stimulation of the frog's auditory system at 800 hertz is not a developmental requirement for normal brain growth and maturation. Indeed, what seems to work best in the anuran species during development is broad-spectrum stimulation across many different frequencies. Hence this particular receiver bias is not a bootstrapping mechanism. Rather, stimulation of the frog's auditory system at this frequency is thought to be related more to detecting potential mates that are, in turn, tuned to vocalizing at this particular frequency.

The types of receiver biases we have been focused on thus far are those linked to the activation of key pleasure circuits in the developing brain. These pleasure-related receiver biases persist into adulthood, when they may play a critical role in driving sexual selection.

As we have seen, sexual selection is often used as a theoretical framework for understanding mate choice, but it goes far beyond this realm in terms of explaining behavioral phenotypes. Evolutionary biologists have never had an easy time accounting for the appearance of so many uniquely human functions such as art, music, humor, and dance from a survival of the fittest perspective. This perspective neglects the obvious fact that our ancestors had to both survive *and* reproduce for their genes to make their way through the ages. Reproduction is itself a competitive act. Individuals must identify what traits are attractive to the opposite sex and do everything possible to amplify their appearance and to hide flaws that might reveal potential weakness. The pioneering biologist Amotz Zahavi argued that organisms are naturally attracted to very specific anatomical features that are used as fitness indicators. A classic example of this is the peacock's elaborate plume. Peahens tend to be attracted to, and prefer mating with, the most highly ornamented peacocks. The question is why. One argument, discussed earlier, is that highly ornamented peacocks are simply more conspicuous to peahens and therefore better at attracting their attention, but careful field studies fail to support this theoretical position. Experiments that remove the impact of this variable by regulating the amount of time different peacocks are exposed to the same peahen still result in a mating preference given to the peacock with the most elaborate plume.

Zahavi and other biologists have taken the position that an elaborate plume signifies the biological fitness of a peacock, since it provides evidence that the animal is strong enough to survive even though the exaggerated plume diverts precious energy resources toward its growth and maintenance. It thus serves as an energy resource handicap that must be overcome. Others have suggested that besides the energy requirement for growth, maintaining an elaborate plume is like wearing a target, since being conspicuous to potential mates also means being conspicuous to predators. In this perspective, the plume indicates fitness because it represents a survival handicap in the context of predation.

The animal kingdom abounds in examples of sexual dimorphisms such as this, a trait that becomes exaggerated in one sex—often the male—and used to attract potential mates. But why do females of the same species develop preferences for these traits in the first place? Humans are clearly not exempt from this, although the exaggerated traits occur prominently in both sexes. From breast implants to the hair weave inspired by a midlife crisis, we spend vast amounts of time and money pursuing activities designed to improve our attractiveness to the opposite sex. However, many of these "improvements" have no obvious impact on our overall health or survival. Sexual selection theory has been used as a theoretical framework for explaining this human propensity for self-adornment. Although there is notable cultural and individual variability in descriptions of physical traits that are preferred in a potential mate, there is also surprising agreement across the globe on what makes a person physically attractive. Given that there is some consensus as to what makes a person attractive, how did these biases toward specific physical features emerge?

The same forms of visual stimuli that play a role in developmental bootstrapping reemerge in the adult as pleasure-inducing receiver biases. Babies who have an innate fondness for faces and strong symmetry grow up to be adults whose eyes linger longest on potential mates with maximally symmetric faces. Small receiver biases that are present at birth can be magnified by ongoing sexual selection. Say, for example, that most females prefer tall men. Even a small bias toward taller-than-average men will have a significant effect on the evolution of male height in generations to come. Given this bias, tall men will have a higher probability of fostering offspring than shorter-than-average men. Assuming both male height and the preference for tall men in women are genetically correlated, their offspring should be taller and, most importantly, prefer taller mates. Hence, the process of sexual selection can take small receiver biases and shape them into widely accepted notions that define physical attractiveness. In the next chapter, we will see how pleasure-inducing receiver biases have become important detectors of fitness during mate selection and hence serve as perhaps the most powerful driving force of sexual selection. This fundamental mechanism has had a profound impact on many facets of our everyday lives.

Part Three

The Pleasure Instinct and the Modern Experience

Chapter 9

Pleasure from Proportion and Symmetry

Our inner faculties are adapted in advance to the features of the world in which we dwell. . . . Our various ways of feeling and thinking have grown to be what they are because of their utility in shaping our reactions on the outer world.

—William James

There must be in our very nature a very radical and widespread tendency to observe beauty, and to value it. No account of the principles of the mind can be at all adequate that passes over so conspicuous a faculty.

—George Santayana

As a boy, the young Charles Darwin showed no signs of the brilliance that was to shine by his twenty-fifth birthday. He and his older brother, Erasmus, grew up playing along the banks of the Severn River in Shropshire, the idyllic countryside setting of Jane Austen's

133

Pride and Prejudice. By most accounts he was an amiable if not astute child who could just as likely be found digging for beetles as attending to the lessons foisted on him by his Latin tutors.

His father, Robert, was a physician and expected the same from both of his sons, who were shipped off to Edinburgh to study medicine in 1825. At the tender age of sixteen, the youthful Charles Darwin was quite shocked by the pace of the city, exposed to a side of life for which he had little conception. The university at the time was the center of a raging debate over Scottish nationalism that seemed an unending battle for both God and country. The greens and lecture halls were filled with rabble-rousers, each coddling their own theological baby, Jacobites, Calvinists, Loyalists, all willing to argue their case. Nor were the faculty, lecturers, and readers immune to this carnival of ideas, some of which proved very dangerous indeed.

Charles learned that his grandfather Erasmus, who died several years before his birth, had somewhat of a cult following at the university. Erasmus was considered an irreverent man by friends and family, and in his later years lived an unconventional lifestyle, championing a free-love movement of sorts. He was trained as a medical doctor and botanist, and his scientific views on nature and religion were even more scandalous than his personal life. In one of his many poems, titled "The Temple of Nature," we find the early seeds of a theory of evolution that his grandson would harvest some sixty years later:

> Organic Life beneath the shoreless waves
> Was born and nurs'd in Ocean's pearly caves;
> First forms minute, unseen by spheric glass,
> Move on the mud, or pierce the watery mass;
> Then as successive generations bloom,
> New powers acquire and larger limbs assume.

Darwin returned home following his second year at Edinburgh, and after fretting for days, gathered enough courage to tell the family he was quitting medical school. He was less at ease with medicine after attending surgical grand rounds and witnessing operations that were then conducted without the benefit of local anesthesia—sights that must have been truly horrific experiences. Robert was outraged by what he took as his son's apathy: "You care for nothing but shooting, dogs, and rat-catching and you will be a disgrace, to yourself and all your family." He decided that if Charles was not to be a doctor, the only respectable alternative was for him to join the clergy. So the young Darwin set off to study theology at Cambridge University, with the aim of understanding the true design of God's nature.

It was in nature that Darwin sought God. He spent much of his free time searching for insects in the forests and fields around the university and reading accounts of archaeological expeditions in South America. He was fascinated by the way other cultures lived and the strange places they called home. An adventurous itch began to build in Charles, and when he was given the opportunity to join Captain Robert FitzRoy and the crew of the *Beagle* on a five-year expedition to map out new trade routes of Argentina and neighboring countries that had just been delivered from Spain's control, he jumped right in. The twenty-two-year-old boarded the *Beagle* on December 7, 1831, a beautiful autumn day in Plymouth, prophetically carrying with him for the voyage a copy of his favorite verse, *Paradise Lost*.

For five years the crew of the *Beagle* charted out the southern waters, past the Canary and Cape Verde islands, on toward Montevideo, Tierra del Fuego, and around Cape Horn. They veered north through the Strait of Magellan, carefully avoiding the iceberg fields that emanate from the Antarctic Circle to within a few hundred miles of the South American coast in spring. Onward, they skirted the shoreline of Chile, stopping for approximately one month at a tiny group of islands known as the Galapagos, just south of the equator.

The time Darwin spent on these islands gathering specimens and the details that he recorded of how functionally well adapted each species seemed to be to its environment had a profound impact on his thinking about evolution, but not until some two years after the visit. While on the islands he recorded a rich variety of birds, particularly the finches, wrens, and warblers that, surprisingly, looked almost identical except for their differing beaks. Some had large, blunt beaks that were used for crushing large seeds, while others had rather narrow and elongated beaks (think needle-nose pliers) that were used to extract small seeds and insects from difficult-to-reach places. Darwin dutifully noted these distinctions and sailed on, returning to England in the fall of 1836 a changed man.

Once home Charles began to receive reports from the leading British zoologists about the specimens he brought back from his travels. These caused confusion at first, but a theory soon emerged that would change the world, and it stemmed from the differences in the birds he saw in the Galapagos. What Darwin failed to realize at the time, but soon learned after returning to England, was that these were *all* finches. The primary factor distinguishing them was beak morphology. It was puzzling how so many different species of finch came to inhabit such a small area, and he reasoned that perhaps they all descended from a common ancestor and gradually, after many generations, began to diverge in appearance. He thought of Lamarck and his "transmutation of acquired characteristics," but he did not accept the idea that changes accumulating within a lifetime could be passed down to offspring. Alternatively, he speculated that the different beak shapes must give each finch a special advantage to living in its local environment. For example, finches with long, narrow beaks would have an advantage in securing food in places that might not be reached by a finch equipped with a larger beak. With a more reliable food supply, the long-beaked finches would have an edge over their natural competitors and be more likely to reproduce. Likewise, in other parts of the islands where elderberry brush was plentiful, finches with more powerful and compact beaks dominate, since they alone can manage to grind the hard casing of their seeds to an edible pulp.

Darwin also knew what farmers understood for years—offspring tend to resemble their parents. Farmers select crops for their next harvest by determining which strains produced the best product this year and replant them hoping to build on their success in future seasons. Eventually this process of farmer selection results in enough accumulated differences that entirely new varieties emerge. In an instant of recognition, Darwin was terrified by the implications of his theory. If he was right, finches (and other animals) are selected by natural competition for food, sex, water, and all means of subsistence, and those that just happen to have an adaptive edge (perhaps because of a longer beak, or keener eyesight, or faster flight) are more likely to reproduce and bear similar offspring, thus ensuring a continuation of that lineage. Other, less successful finches that do not possess the advantages are less likely to survive to reproductive age in that environment, thus minimizing the spread of their characteristics to offspring. The different varieties of finches he saw on the islands were functional success stories—the good seeds. Each had some adaptations that secured reproductive success in their given environment at the expense of competing birds.

This was a deeply sad story for Charles, a religious man who had great difficulty accepting the idea that finches and other birds—indeed, all animals including humans—must evolve over time. Animal forms are not fixed through eternity, but rather molded by a process of natural selection—imagine, selection of a species without a Selector. It took Darwin fourteen years of confidence-building before he made his theory public in the magnum opus *On the Origin of Species by Means of Natural Selection.* The public outcry of blasphemy was, as expected, enormous, and included many of his scientific colleagues, friends, and even family (his wife wasn't thrilled with the theory). Of course, an unwillingness to accept species selection without divine intervention continues today.

Darwin concentrated on evolution by natural selection in *Origin of Species*, which made virtually no reference to human behavior. This is not an oversight. Darwin was troubled by the lack of an obvious way to account for the development of so many uniquely human activities, such as making music and art, by the theory of natural selection. What is the survival function of singing a pretty melody, making someone laugh, being able to tell a good story, or creating a work of art? None of these human qualities seemed to fit into the same theory as that of the finches growing different-size beaks to adapt to distinct environmental challenges.

In his follow-up work *The Descent of Man and Selection in Relation to Sex*, Darwin reconciles the appearance of these behaviors by developing the theory of evolution by sexual selection—a special form of natural selection. The bottom line of evolution is the survival of genes down through the generations. An organism may survive to a ripe old age, but if it fails to reproduce, its genes die right along with it. For selection to drive evolution, an organism must survive and reproduce.

The adaptations that result from sexual selection are referred to as ornaments by evolutionary biologists. The example given earlier was of the peacock's enormous and well-decorated plumage, which plays a direct role in attracting the attention of peahens. Darwin argued that the evolution of this display was driven by female choice.

In species where there is competition for mate selection, the elaboration of secondary sexual characteristics (that is, those not serving a direct function in reproduction) usually occur in males to vie for the attention of females, who have a greater metabolic investment in reproduction.

The English geneticist Angus Bateman observed that in many species, females bear a much larger burden for producing an offspring than males. This inequity begins with the production of sex cells. Women produce approximately four hundred nutrient-rich ova during an entire lifetime, while men produce billions of sperm that are replenished at a rate of about eleven million to twelve million per hour. A female has a relatively small number of eggs at any given time, and a single male can fertilize all of them; hence the female will not produce more offspring by mating with more than one male. Contrasting this, males are capable of fathering many more offspring than any one female can bear if he mates with several at a time. Of course, fertilization and gestation occur internally within females, consuming great amounts of time and metabolic resources. If a successful birth occurs, this is followed by potentially several years of lactation to feed the child and still many additional years of investment in raising the toddler to an autonomous age. Thus, in many species where these conditions exist, females must be far more choosy in selecting mates than males.

Female choice has been shown to drive a broad range of adaptive traits or ornaments in males. A key question that has troubled biologists for decades is how particular traits are selected in the first place. For a trait to form through sexual selection there must be some initial preference for it by the female. Early theorists such as Sir Ronald Fisher, a geneticist who also made pioneering contributions to modern statistics, argued that the initial preference might be completely arbitrary. Say, for example, that a small number of female peahens developed a preference for mating with peacocks with brighter-than-usual plumage. Peacocks having brighter plumage would be more likely to mate with these peahens and produce peacock offspring with brighter plumes and peahens with a preference for bright plumes. Assuming the traits (producing the bright plume and preferring the bright plume) become genetically correlated, this would produce a positive feedback system where both traits become increasingly magnified over many generations until plumes become so large that they begin to have survival costs associated with their production that temper the process. At this point, greater elaboration of the plume should be curtailed, since it will result in survival deficits. An equilibrium point would emerge where the positive sexual selection effects of large, bright plumes are perfectly balanced by their negative survival costs.

But why would peahens develop a preference for larger, brighter plumes in the first place? We can find some help here in the work of biologists such as William D. Hamilton and Marlena Zuk, who pointed out that many ornaments are excellent indicators of genetic fitness. The fitness indicator theory suggests that any population where there is an imbalance of investment in producing offspring should theoretically result in the sex with the greater investment also having more at stake in mate selection. This would create pressure for that sex to be able to identify and pair with the most genetically fit mate available. In this example, female choice involves finding the best male genes with which to join with her own so that her offspring have the best chance of survival and reproducing. Of course, finding the best genes is tricky.

One argument might be that the best male genes are those that are fairly different from the female's own, since this will reduce the possibility of recessive combinations being expressed (genetic diversity prevailing over homogeneity). Another argument might be to simply find genes that are healthy in general. Either way, it has been shown that in many sexually reproducing animals (including humans), key traits that are reliably associated with genetic fitness can be detected and used for mate selection.

The fitness indicator theory of sexual selection is fairly convincing, since it explains a great deal of empirical observations across many species where the development of a particular trait seems to have extended well beyond the limits imposed by survival costs. The case of the peacock's plume is a perfect example in this regard, since its exaggerated growth makes it a target for predators. Indeed, the handicap principle championed by biologist Amotz Zahavi (see chapter 8) says that it is exactly this cost that makes it a relevant fitness indicator.

Fitness indicator theory goes a long way in explaining why some traits seem to be taken to extremes beyond which survival costs would be accrued. It also helps explain how initial female preferences for a trait, if reliably associated with genetic fitness, can emerge. Peahens that have preferences for traits that have poor or no correlation with genetic fitness are in trouble. If they use these traits for mate selection, they are essentially gambling with their reproductive success. If they have chosen poorly, their genes, along with the trait preferences they support, will be less likely to thrive and be propagated.

In this context it is easy to imagine how pleasure might play a pivotal role in this process. As we have seen in the chapters to this point, the pleasure instinct drives the emergence of distinct receiver biases—preferences for certain forms of sensory stimulation that are critical for normal brain development and maturation. If pleasure-associated preferences for particular forms of stimulation guide an organism toward traits that are also good fitness indicators, this combination may prove to be very useful during mate selection.

Imagine Sally is being pursued by Harry. If Sally makes her selection of a mate purely on the basis of fitness indicators without regard to whether they bring her pleasure, she might do so by simply summing up the tally and attributing an equal value to each indicator. A completely different approach would be to use the pleasure associated with the appearance of a particular fitness indicator to gauge its current importance relative to others. In this context, pleasure is the common currency for organizing and prioritizing competing goals and interests. Using this approach, Sally would be able to assess and rank-order which fitness indicators are more important than others so she could make a more informed choice that matches her current needs. This process would be far more flexible and adaptive to changing environmental circumstances than one based on a simple summation of overall values across all indicators.

If, at this particular time, Sally has not eaten for two days and is faced with a choice between Harry, who has taken her to dinner, and Tom, who has taken her to a movie sans dinner, the choice will be very different depending on the approach. If she chooses solely on the basis of fitness indicators, either Harry or Tom will do, since they have both done things to display their fitness. Harry has shown that he can provide food, a critical resource for survival. Tom has shown that he has sufficient wealth that he can waste money on things not directly related to survival. Both have provided evidence of their fitness. If forced to choose simply on the basis of fitness indicators, Sally might have to toss a coin to decide. A different outcome would occur, however, if pleasure is used to rank-order the relative importance of fitness indicators to match her current needs. In this scenario, Harry is the clear winner, since he has delivered what she needs most at present—nutrition.

Seen in this light, pleasure provides the common reinforcement mechanism to drive and align motivated behaviors that may require very different forms of learning and that likely occur in different sensory systems. Pleasure is nature's shortcut; it enables humans to respond quickly to changing life demands by prioritizing basic needs that involve different neural systems on a single metric. The brain's motivational systems for signaling hunger for food and hunger for sex are distinctly different. Both brain systems, however, interact with the brain-stem pleasure circuitry we have been discussing. The pleasure system, in this respect, is the common denominator that allows a direct comparison of the needs associated with both competing hungers so that appropriate behavior can be chosen based on current priorities.

What are some examples of receiver biases (see chapter 8) crafted by the pleasure instinct that are also good fitness indicators? Certainly not all fitness indicators are consistent with the receiver biases created by the pleasure instinct. Likewise, most of the receiver biases we have discussed thus far are good fitness indicators, although there are, of course, exceptions. Let's look at a few classic archetypes to guide our thinking. We will see that such biases occur in both sexes in humans, since mating generally involves monogamous pair bonding.

The Hidden Persuaders

It is clearly not a controversial claim to suggest that different cultures and different times tend to produce variable ideas of what is most physically attractive in a potential mate. Civilizations from the ancient Greeks onward have attempted to formulate canons for defining the ideal of physical beauty. Plato and Plotinus wrote extensively about the geometry of physical form and emphasized the inherent aesthetic appeal of things that exhibit strong symmetry, harmonious proportion, and vivid color. The emphasis on symmetry and proportion—elements that are quantifiable—began in the fifth century B.C. and has been built upon steadily by artists and philosophers ever since.

During the Renaissance, an explosion of aesthetic theories applied to human forms emerged, many of which concentrated on identifying the proper metric to measure true beauty. Leonardo da Vinci and Albrecht Dürer were among the most popular artists who proposed geometric systems for measuring beauty based on symmetry and proportionality of body parts. During this period, a large number of measurement systems were proposed, each one emphasizing a particular set of metrics. For instance, Dürer, inspired by the great Italian artist Jocopo di Barbari, created a formal theory of beauty based on anatomical proportions such as finding correspondence between finger length and palm width, arm length to an even-integer ratio of body length, and so forth. Although some artists from this period, such as Leon Battista Alberti, believed there was a single irreducible geometric form representing perfect beauty, da Vinci and Dürer were more accepting of relative beauty in that many forms could be seen as being equally beautiful provided a few basic ratios were preserved.

The canonization of beauty continues today in modern attempts to determine if a universal definition can be formulated that is consistent across cultures. While it has proven difficult to show that any of the historical canons match up with what modern people actually find attractive, there have been interesting findings from some of these studies.

What we generally find in modern studies is that there is no ideal physical form that all will agree on as being beautiful based on pure mathematical principles—no golden ratio of beauty. However, there do seem to be certain physical traits that people from widely diverse cultures (Western, Middle Eastern, Eastern, north and south of the equator) agree on as being beautiful or attractive in a potential mate. The questions are: What are they? Why do we like these traits so much?

Pleasure from Proportion

Let's start with generic body form. The most obvious quantifiable traits that characterize body appearance are weight and height. When one looks just within a single culture, it is easy to find that certain height and weight ranges are thought to be more attractive than others. In parts of North America and Europe, there tends to be a preference for tall and thin models (of both sexes), while in some South American and Polynesian cultures, those with a little more weight are considered most attractive. Given this variability for ideal height and weight by different cultures, these are clearly not universally accepted traits for beauty. A young woman raised in the Bronx might look at a possible suitor who is tall, dark, and handsome with a winsome eye, yet the same man might be seen as meek and too skinny for the likes of a highlander from Papua New Guinea. Indeed, the man might be seen as meek and too skinny for the likes of another Bronx native. Barring extremes, it turns out that body weight and height are fairly poor predictors of whether an individual is generally found to be attractive.

What seems to matter most in determining attractiveness is the overall body shape of a person. People of the same height and weight can have bodies that look remarkably different. Body shape is driven by the distribution of body fat, and as we will see, this trait is significantly correlated with a woman's sex hormone profile, reproductive capability, and risk of disease. In humans, the distribution of body fat depends on both age and gender. Boys and girls have strikingly similar distributions in infancy and early childhood. At puberty, hormonal changes lead to a shifting of these distributions. Increased estrogen in postpubertal girls blocks fat buildup in the abdomen and stimulates buildup in the buttocks and thighs. Increased testosterone in postpubertal boys does quite the opposite, causing increased fat deposition in the abdomen and decreased buildup in the buttocks and thighs. In general, women have greater amounts of fat in the lower parts of the body (gynoid or pear shape), and men have greater amounts in the upper portions of the body (android or apple shape).

Differences in body shape can be reliably quantified by measuring the circumference of the waist and hips, and calculating a waist-to-hip ratio (WHR). Male and female prepubertal children have similar WHRs. After puberty, gender-specific hormonal changes shift fat distributions within each sex such that a woman's WHR tends to be smaller than that of a man. Healthy premenopausal women typically have a WHR from .67 to .80, whereas healthy men usually have WHRs between .85 and .95. Hence, in women, WHR can be used as a reliable means to gauge an individual's general reproductive status (pre- versus postpubertal).

In the early 1990s, psychologist Devendra Singh of the University of Texas began publishing a series of papers demonstrating that men and women from very different cultures display remarkable similarity in what WHR they find attractive in a female. You may find it odd that Singh used both males and females in his studies of female attractiveness, but this is a critical control from an evolutionary perspective. If a trait is to be a reliable marker of attractiveness, both the signal receivers and the signal generators must be aware of its meaning. Woman, realizing that a particular trait is seen as attractive by men, might wish to accentuate or attenuate it by various means (for example, makeup, clothes, posture, and so forth) based on the desire to indicate sexual availability.

In his first series of studies, Devendra Singh created a set of line drawings depicting women with three body weight categories (underweight, average, and overweight). Within each weight category, he used line drawings to represent four WHRs. Two were typical gynoid shape with WHRs of .7 and .8, and the other two had a typical android shape with WHRs of .9 and 1.0. He and his colleagues showed the drawings to men and women of different ages (eighteen to eighty-five years old), professions, educations, and ethnicities, and asked them to rate each figure based on its attractiveness. The results were very interesting. Men and women rated drawings with a WHR of .7 as the most attractive within each weight category. The drawing seen as most attractive was of a female figure with a WHR of .7 and average weight. The drawings seen as least attractive were of female figures with .9 and 1.0 WHRs from the overweight category.

A possible critique of these findings might be that they have focused entirely on Western cultures, since Singh's studies only included Americans of European, Mexican, and African descent. To show that a preference for a particular trait is shaped by a selection process, the first step is to demonstrate that it is at work in humans of diverse cultures and ethnic groups. To see if WHR is indeed a universal marker of attractiveness, Singh next did a series of studies using men and women from nineteen different cultures. In these studies, he included people from America, Europe, Australia, Africa (Kenya, Uganda, Guinea-Bissau), the Azore Islands, the Shiwiar tribe of East Ecuador, Indonesia, China, India (Sugali and Yanadi tribes) Chile, and Jamaica. He showed the same line drawings as used in the earlier studies and asked subjects to rate their attractiveness, healthiness, youthfulness, and desirability as a marriage partner.

Despite coming from incredibly diverse cultural backgrounds, there was a clear preference for a WHR of .7 in each weight class. Moreover, there was a strong correlation among the variables in the findings. Perceived attractiveness, youthfulness, and healthiness were ranked in an almost identical fashion across the different cultures. For these variables, positive rankings decreased systematically with increased WHR.

One might argue that there could still be the influence of Western media on these findings, since it tends to associate a particular body shape with beauty. Granted, but this is unlikely to have impacted these results, since both the Azoreans and subjects from Guinea-Bissau had virtually no exposure to Western media, yet they ranked the drawings similarly to most other cultures, including those from the United States. Hence it seems that WHR may be a universal marker for attractiveness in a potential mate. In females, the maximal ranking seems to be about .7. Of course, as mentioned earlier, many different body types and images of beauty can have a WHR of .7. The classic beauties Marilyn Monroe at 36–24–34 and Audrey Hepburn at 31.5–22–31 had very different hourglass figures, but shared the same WHR of .7.

The next logical question to ask, given the seemingly universal appeal of a .7 WHR in females, is why such a trait is important. Why do we find looking at a female with a .7 WHR more pleasurable than one with a WHR of, say, 1.0? The answer is not to be found in some ancient canon of beauty, prescribed by the divine or mathematical. Rather, it can be found in the way WHR reveals basic information about the bearer's general health and fecundity. Reproductive success for a man in ancient environments must surely have depended on selecting a mate with good health, genetic fitness, and excellent reproductive capacity. Of course, these characteristics are not directly observable; hence sexual selection has shaped certain mental mechanisms for measuring genetic fitness indirectly. There is now convincing evidence demonstrating that WHR is a fairly good predictor of long-term health risk, mortality, and reproductive endocrinological status.

Women with a WHR lower than .8 have a significantly reduced risk relative to women with a WHR above .8 for key conditions that are known to hinder reproductive success and fertility, including hyperandrogynism, menstrual irregularity, suboptimal sex hormone profiles (optimal is high estrogen and low testosterone), and abnormal endocervical mucus pH. Women with WHRs below .8 are also significantly more likely than their age-matched counterparts with WHR above .8 to have a successful pregnancy outcome after artificial insemination or in vitro fertilized embryonic transfer. Thus WHR is a reliable marker for estimating a woman's reproductive health.

It is important to note that these are statistical observations. Certainly, there are many women with WHR well above .8 who have perfect reproductive and general health. Likewise, there are undoubtedly loads of women with WHR of .8 and below who do not share this health. Sexual selection shapes traits that are expressed to various degrees within a population. It is nature's way of playing the odds. If a healthy man mates with a female with a WHR of .7, there is no guarantee of offspring. Sexual selection has crafted a psychological mechanism—a preference for females with a particular WHR range—that, all other things being equal, increases the odds of producing offspring who will preserve his genes, since the female has better odds of increased genetic fitness, disease resistance, and fecundity.

Seduction by Symmetry

Another excellent example of a receiver bias crafted by the pleasure instinct that does double duty as a fitness indicator is our love of symmetry. In his book *The Descent of Man and Selection in Relation to*

Sex, Charles Darwin wrote, "The eye prefers symmetry or figures with some regularity. Patterns of this kind are employed by even the lowest savages as ornaments; and they have been developed through sexual selection for the adornment of some animals." As we saw in the previous chapter, at about the time the primary visual cortex (V1) reaches maximal cell proliferation, babies begin to become keenly attracted to objects exhibiting strong lateral symmetry. As if on cue, this preference emerges just at the right time to promote experience-expectant synaptic pruning of V1 and downstream visual cortical areas, hence promoting normal maturational development. The pleasure obtained in self-stimulating V1 and downstream visual areas with highly symmetric objects as an infant forms the basis for a host of preferences as adults, including an attraction to things bearing strong lateral, rotational, and radial symmetry. The most obvious example of this is how this preference impacts our choice of mates.

As we have been discussing, the fitness indicator theory of sexual selection suggests that individuals develop preferences for potential mates who possess traits that are reliably linked with good genetic quality and increased likelihood of bearing viable and vigorous offspring. Indeed, many sexually selected traits (such as WHR) seem to have comparable genetic variability to that commonly observed for fitness traits (for example, fecundity), which indicates they may have evolved as signals of overall phenotypic condition.

One important marker of phenotypic condition that has caught the attention of researchers in the past decade is bilateral symmetry of the body. Imagine a line drawn from top to bottom down the middle of your body. If I were to take calipers and carefully measure the relative size of many body parts (such as the width of your feet, ankles, and ears, or the length of your fingers from both sides), I would typically find slight variations in relative size from left to right. The variations are usually small, on the order of about 1 percent of the overall size of the body part being measured. The distribution of these bilateral differences is referred to as fluctuating asymmetry, since departures from perfect symmetry vary randomly along the body axis. Fluctuating asymmetries are very different from directed asymmetry, such as handedness, since the former average out to zero in the general population.

Fluctuating asymmetry can be caused by a number of factors during development. Since the corresponding sides of the same body part are coded by the same genes, fluctuating asymmetries typically emerge from either environmental stressors or genetic perturbations within the genome that reduce developmental stability. Such stressors include things such as parasites, pathogens, pollutants, and other environmental challenges such as extreme temperatures or marginal habitats. Fluctuating asymmetries also increase with genetic perturbations caused by things such as inbreeding, the presence of certain recessive genes, chromosomal abnormalities, and homozygosity. Considering this, it is thought that fluctuating asymmetry is a measure of the extent to which an individual has been able to maintain a normal developmental trajectory by resisting such challenges. A person with a large number of harmful genetic mutations or who is less able to resist pathogens should on average exhibit greater fluctuating asymmetry.

There is now a substantial body of literature showing that fluctuating asymmetry is a reliable indicator of overall phenotypic quality. Increases are associated with decrements in biological fitness in a number of key domains, including reproductive success, growth rate, an ability to resist disease, metabolic efficiency, immunocompetence, and overall survival rate. Insofar as fluctuating asymmetry has been found to be partly heritable and is a reliable marker of phenotypic quality and biological fitness, several authors have suggested that ultimately it is a marker of genetic quality. As such, theoretical models of sexual selection by mate choice and competition would predict that fluctuating asymmetry should exhibit a strong relationship with mating success. That is, individuals with increased symmetry should, in general, enjoy more successful mating than their more asymmetric counterparts.

Indeed, in the majority of species tested, males with the highest degree of bilateral symmetry tend to have the greatest mating success. In a large-scale review of sixty-five studies involving forty-two species across four major taxa, biologists Anders Moller and Randy Thornhill found a number of interesting results that were predicted by theoretical models years earlier.

First, in the vast majority of species tested—from fruit flies to humans—males showed the strongest association between fluctuating symmetry and measures of mating success. Females also exhibited a statistically significant relationship between fluctuating asymmetry and mating success in many species (especially humans), but the relationship was strongest in males across most species, which is exactly what would be predicted from evolutionary models of sexual selection based on female choice. Choosier females would lead to greater sexual selection in males, who would, in turn, develop more pronounced traits (for example, ornamentation) to be assessed by females as part of their selection criteria. Accordingly, a sexually selected trait such as fluctuating asymmetry should bear a stronger relationship to mating success in males who have to compete with each other for female attention in species where female choice dominates.

In contrast, there should be a more pronounced relationship between fluctuating asymmetry and mating success for females in species where male choice dominates. In species where mate choice is more equitable among the sexes, the relationship between fluctuating asymmetry and mating success should exist in both sexes to roughly equal amounts.

A second important finding was that in most species, including humans, the relationship between fluctuating asymmetry and mating success was stronger for body parts involving secondary sexual characteristics (traits that distinguish the two sexes of a species but that are not directly part of the reproductive system) than for other parts. For instance, in the dozen or so studies of humans, investigators have examined symmetry at several stops along the primary axis of the body, including the feet, ankles, hands, fingers, arms, chest, shoulders, ears, face, breasts, and the overall figure. In terms of gauging mating success, researchers have measured things such as the rated attractiveness of a potential mate, the likelihood of accepting a date with that person, the likelihood of engaging in sex with them, and others. In general, the most persistent and robust relationship between fluctuating asymmetry and mating success in humans was found to involve parts of the body that are most meaningful during intimate encounters, such as the face, shoulders, chest, and breasts. Why should this be the case? One might argue, perhaps, that it is exactly features such as the face, chest, and breasts that naturally draw our attention because we find them pleasurable, so we are inclined to tune in to them when deciding about a possible mate. But this is a circular argument in this context. True, we focus on these features because we find viewing them pleasurable—certainly more so than focusing on a potential mate's ear length or ankle width. But the primary question is: Why is it so much fun to look at these features in the first place? A second, related question is: Why do these features carry more relevant information (in terms of fluctuating asymmetry) than, say, our feet, knees, or elbows when gauging the suitability of a sexual partner?

When considered from the "good genes" perspective, it would seem that symmetry of secondary sexual characteristics such as the face and breasts varies most strongly with mating success because these features are hard to fake (at least in ancestral times) and are honest indicators of true fitness. For instance, the biological complexity and metabolic cost of building a face makes it particularly susceptible to genetic or environmental perturbations during development that would leave a visual record of such events in the form of increased asymmetry. As discussed earlier, preferred features that are genuinely related to fitness should increase in prevalence, assuming that the preference and appearance of the trait are genetically correlated. This increase in the expression of the trait is tempered by metabolic costs associated with its development. Without such constraints, a simple Fisherian model would predict a runaway process of the trait becoming ever exaggerated and everyone having maximal symmetry throughout the population. This obviously does not occur. Body parts that are the most costly to build are the best candidates for being honest fitness indicators, since they have developed despite metabolic costs and the possibility of environmental or genetic perturbations.

Another important reason why symmetry of secondary sexual characteristics might be more important indicators than symmetry of other body parts is that these features change dramatically at puberty, announcing sexual maturity. Take the face, for example. It is often difficult to determine the sex of a baby by just looking at its face if no supporting clues are available such as gender-typical clothing. Toddlers can also have very similar facial appearances across the sexes, but marked differences in facial appearance generally occur by puberty. During adolescence hormonal changes sculpt these differences. Boys' faces become larger and more angular, especially the lower jaw and brow ridge. Girls' faces retain a smoother forehead and smaller lower jaw, giving a rounder impression. A smaller nasal bridge relative to boys gives the impression of larger and wider-spaced eyes in girls. Clearly, the developmental growth of a face involves much more than simply scaling up the size of the prepubertal face.

Many body parts undergo a qualitative change at puberty where fat redistributes in a sex-specific manner. By comparison, however, the face undergoes extreme changes with many more opportunities for environmental and/or genetic challenges to the developmental stability needed to achieve perfect symmetry. This sensitivity to challenges that occur during development is what makes facial symmetry a potential selection mechanism for identifying mates, assuming it correlates with actual fitness. So let's see what is so important about faces that make them the center of attention in the mating game.

A Fit Face

As we saw in earlier chapters, there is something special about faces that makes them a naturally pleasurable stimulus. Even babies that are a mere ten minutes old gaze longest (a proxy for measuring preference in infants) at illustrations with anatomically correct compositions of a face when compared to control illustrations that have all the same components but are ordered in a random manner. They will also visually track a line drawing of a face at this age. Right out of the womb, babies have a preference for faces. Within another day, newborns develop a preference for their mother's face as opposed to that of other similarly aged women who have recently given birth. By day three, infants can mimic certain facial expressions, such as sticking out a tongue in response to a similar gesture from Mom or Dad. Add a few months and infants develop an ability to discriminate one unfamiliar face from another and detect different emotional expressions, of which they prefer joyful over angry faces.

There might be many different reasons why faces seem naturally interesting and attention-grabbing to humans. The prevailing theory is that an infant's fascination with faces emerges as an adaptive mechanism to promote parent-child attachment. Being able to recognize and engage the primary caregiver increases the likelihood that an infant will become emotionally bonded with that individual and receive proper nurturance. The need to recognize, engage, and extract information from faces continues, of course, through childhood and into adulthood. Being able to read the minds of others in a social group is also important for survival and reproductive success. Humans can't read minds, but the next best thing is being able to understand the emotional mind-set of your peers. No other body part even comes close to yielding such rich emotional information about the bearer as is the case with the face.

The "face as a kin recognition device" theory is well supported in the literature and is consistent with a large number of primate studies, including humans. There may, however, be additional reasons why infants (and adults) find faces so pleasurable. Faces are composite objects made up of smaller, complex stimuli—eyes, lips, nose, jaw, brows, skin, and so forth. Each of these elemental objects is itself a potentially rich source of stimulation for growing brains and indeed has physical characteristics that are known to be naturally preferred at or near birth by newborns. For instance, newborns have a preference for stimuli with strong lateral symmetry and can recognize vertically symmetric objects (symmetric around a vertical axis like the letters "A" and "V") more quickly than asymmetric objects. They also prefer objects that are smooth rather than rough, complex rather than simple, and have high-contrast contours, curves, and concentricity. Faces generally have smooth skin punctuated by high-contrast elemental objects. The elemental objects all exhibit strong lateral symmetry, high-contrast curves, and a high degree of concentricity (as does the whole face).

In this respect, the face might be regarded as a veritable treasure trove of pleasure-inducing stimuli. For the newborn, even an unfamiliar face is pleasure-inducing because it stimulates multiple core features across multiple sensory domains (for example, touch, audition, and vision) that its experience-expectant brain requires for normal development. Clearly, the pleasure experienced by a newborn looking into a caregiver's face only increases over time as it comes to recognize this person and bond with her or him. Before this occurs, however, there must be a neural mechanism that increases the likelihood that the newborn spends more time looking at faces rather than, say, knees. The face provides a hedonic wonderland for the newborn, since a single experience can stimulate developing visual, somatic (touch), and auditory cortical regions in an integrated manner. As we saw in chapter 7, newborns also prefer certain forms of auditory stimulation, such as sounds that have slowly increasing and decreasing pitch contours—the singsong melody of motherese. Motherese, of course, emanates from mouths embedded in faces. Such pleasurable sounds serve to draw the attention of the newborn to the face, where they can be grouped together with other pleasurable features.

Not only do infants find faces pleasurable, they also differentiate attractive versus unattractive faces in much the same way as adults. In a series of compelling studies, psychologist Judith Langlois and her colleagues at the University of Texas at Austin found that infants as young as two months old prefer to look at attractive faces more than unattractive faces as rated independently by adults. Langlois and her colleagues began their study by taking a large collection of color slides of female faces to groups of undergraduate men and women and asking them to rank each slide based on its attractiveness from

1 (least attractive) to 5 (most attractive). The female faces were posed with a neutral emotional expression and glasses removed. Moreover, all clothing was masked so that judgments could be made based on facial features alone. There was remarkably high agreement across raters on attractiveness (0.97 coefficient alpha). Two groups were formed based on the rankings, eight slides with the highest attractiveness rating and eight slides with the lowest rating.

To determine preferences, infants were seated in their mothers' laps and shown a series of two slides positioned side by side. One slide was from the most attractive group, the other from the least attractive group. Recordings were made of the amount of time each infant spent looking at the different faces. Differential fixation time is a common metric for evaluating preferences in infants. This metric generalizes in that children and adults also look longer at self- and independently rated attractive faces than at unattractive faces. Langlois found that infants spent significantly more time gazing at faces from the most attractive group than at faces from the least attractive group. This result has since been replicated and extended to show that infants also prefer more attractive faces to lesser attractive faces within other groups, including both male and female adult Caucasians, male adult African Americans, and other infants.

Needless to say, this finding was not consistent with the prevailing theories of child development at the time. Before Langlois began her research, many researchers assumed that preferences for attractiveness were based on the gradual learning of standards within a culture through a variety of sources (for example, media and social experiences) and that these emerged much later during development. Her data are at odds with this view, since these infant subjects have presumably had very limited exposure to such forces, in that they are preverbal and only two months old. Rather, these findings suggest that preferences for attractiveness are either in place at birth or shortly thereafter.

At present, many studies have shown remarkable agreement across raters from different age, gender, and cultural groups in terms of facial attractiveness rankings. Cross-cultural studies have been done with people in the United States, throughout Europe, China, Korea, South America, and Asia using multicultural faces, with fairly consistent agreement on which faces are the most attractive. A major question that is the focus of much investigation is what features humans use to make determinations about attractiveness and why,

Studies by evolutionary biologists and psychologists have found that people sometimes find the average attractive. Take a thousand faces, average their spatial characteristics, and you get a new face representing the group norm, which is generally rated to be slightly above average in attractiveness. In his classic text on the evolution of human sexuality, anthropologist Donald Symons hypothesized that natural selection drives adaptations, where at the population level, the optimum value of some trait is likely to be the mean. In this respect, a preference for the average value of a trait such as facial characteristics would be wise, since it would push the chooser toward mates with optimally adapted facial traits for things such as breathing, chewing, and any other functions linked genetically to the development of facial features. Randy Thornhill and psychologist Steve Gangestad theorized that a preference for averageness occurs because for some heritable traits, the mean value represents maximal genetic heterozygosity (allelic diversity).

Other studies have found that people tend to focus on secondary sex characteristics when making judgments of attractiveness. In humans and many other species, hormonal changes occurring at puberty can actually handicap an individual. For instance, increased testosterone production in adolescent males leads to increased musculature and energy expenditure. These changes, in turn, raise the metabolic demands of the body and draw resources away from other systems, including immune functioning. Since testosterone production creates a draw on immune function, it may be less costly for genetically fit males to have high levels of testosterone than for those more susceptible to environmental and/or genetic perturbations (for example, pathogens, maladaptive environmental conditions, genetic mutations). Likewise, increased estrogen production in pubertal girls indicates reproductive potential and fertility. Reproductive effort places increased metabolic demands on a woman's body and draws resources away from other important biological processes, including immune function. Hence, secondary sexual characteristics of the face that become exaggerated at puberty (such as an extension of the lower jaw and broadening of the brow ridge in males) may be honest signals of phenotypic and genotypic condition.

In addition to averageness and attention to secondary sexual characteristics, many studies have found a strong relationship (perhaps not surprisingly) between ratings of facial attractiveness and fluctuating asymmetry of the face. Indeed, as mentioned earlier, in their extensive review, Anders Moller and Randy Thornhill found that in most species, the correlation between fluctuating asymmetry and mating success was strongest for body parts that are secondary sexual characteristics. The face is one such body part, since, as we have seen, it is replete with secondary sexual characteristics that become differentiated at puberty.

Numerous studies have demonstrated that both facial and body symmetry involving secondary sexual characteristics are related to reproductive success and health in general. In a diverse range of species tested, increased fluctuating asymmetry has been shown to be related to decreased fecundity, growth rate, survival, and metabolic efficiency. For instance, increased fluctuating asymmetry in men has been shown to be related to a number of fertility measures. Population biologist John Manning and his colleagues from the University of Liverpool studied males referred to a reproductive medicine clinic at a local hospital for routine semen analysis. They found that men with greater body asymmetry had fewer numbers of sperm per ejaculate, lower sperm speed, and reduced sperm migration relative to their more symmetric counterparts. In women, breast asymmetry has been found to correlate negatively with fecundity and to the probability of marriage.

Other studies have shown that body symmetry is related to how efficiently our bodies use energy. If one has a normally low metabolic cost associated with the maintenance of body processes, this should theoretically free energy for use in other ways, such as maximizing developmental homeostasis. Consistent with this ideal, Manning's group has also found that people with greater body symmetry exhibit a lower resting metabolic rate, measured as oxygen consumption at rest.

The relationship between overall physiological health and body symmetry has also been examined by several investigators. Gangestad and Thornhill gave a series of health questionnaires to 203 romantically involved couples where both the man and the woman rated self and partner. The questions asked about the individual's current condition in eight specific physical health domains, including muscularity, energy, stamina, vigorousness, robustness, lethargy, physical tightness, and cardiovascular fitness. The researchers found a significant negative correlation between a composite physicality score (summed from the individual domain scores) and body asymmetry. As asymmetry increased, perceived general physical health decreased.

Other investigators have found similar results when focusing just on facial asymmetry. In a study of 101 college students, evolutionary psychologists Todd Shackleford and Randy Larsen examined the relationship between facial asymmetry and a broad of range of physical, emotional, and psychological health indicators. Head shots of students were taken and used to extract estimates of facial asymmetry using bilateral measurements at the outer eye, inner eye, nostril width, cheekbone width, and jaw width. Each subject was also given an extensive battery of health questionnaires concerning mood and emotional state, personality, life orientation, sociability, impulsivity, and general symptomatology. Additionally, subjects were requested to complete reports on their daily activities, moods, and physical symptoms twice each day. These included standard physical symptoms such as headaches, trouble concentrating, runny nose, sore throats and coughs, gastrointestinal problems, and so forth. Finally, each participant had his or her overall aerobic fitness assessed by measuring cardiac recovery time following a standard exercise protocol to raise their heart rates to at least thirty beats per minute beyond resting state.

A separate set of college-age observers were asked to rate each photograph of the subjects along a number of dimensions including attractiveness, happiness, reliability, agreeableness, intelligence, emotional stability, activeness, and others. The overall results of the study painted a complex picture revealing how variation in facial asymmetry impacts several key aspects of daily life in the bearer and how others view them.

In terms of physical health, subjects with greater facial asymmetry were more likely than their symmetric counterparts to report negative physical symptoms, such as backache, muscle soreness, reduced vigor, and trouble concentrating. They were also more likely to complain of depression, perform more impulsive acts, and view their lives as being outside of their personal control. Men with greater facial asymmetry, in particular, tended to score higher on measures of mania and schizophrenia components of the Minnesota Multiphasic Personality Inventory (MMPI), a classic clinical assessment, than those who were more symmetric. Consistent with this observation, several studies have now shown that schizophrenia in adults and hyperactivity disorder in boys are both associated with increased body and facial fluctuating asymmetry.

On the flip side, individuals with relatively more facial symmetry when compared to their counterparts were more likely to be optimistic, view themselves as superior, and score higher on measures of narcissism. Interestingly, extroversion seems to correlate positively with the degree of facial asymmetry in women and negatively in men. Women with more facial asymmetry tend to be more extroverted, while men tend to be more introverted.

Independent judgments by external observers were also systematically related to the degree of facial asymmetry in the subjects. Individuals with greater facial symmetry were viewed by others as being more conscientious, intelligent, active, agreeable, and genuine than those with less symmetry. This study confirms other reports that facial asymmetry correlates with a number of important markers of physiological, emotional, and psychological health.

Symmetry Signals and Pleasure

Thus far we have found that increased body and facial asymmetry are associated with decreases in certain indicators of fertility, as well as self-reported physical, emotional, and psychological well-being. But does greater facial symmetry translate into improved mating success? In other words, do people actually have a preference for symmetric mates?

Symmetry is a basic property that is preferred and sought after by newborns and toddlers alike. They recognize more quickly and exhibit greater pleasure in viewing symmetric objects than those that are asymmetric. I have shown in earlier chapters that an anatomical and developmental imperative exists for symmetry-seeking almost immediately after birth, since it represents an optimal form of stimulation for experience-expectant maturation of the primary visual cortex and additional downstream visual areas (for example, V2, V3, and inferotemporal cortex).

The preference for symmetry right after birth would be expected to continue through childhood, since synaptic pruning of these visual cortical areas continues for decades. This early preference thus forms the basis for a fondness of symmetry in adults. Moreover, since body and facial symmetry are related to general health (and are, hence, potential fitness indicators), an inborn preference for symmetry that mediates normal brain growth and development would be a perfect trait to be co-opted through sexual selection mechanisms. If this scenario is true, one would predict all six of the following conditions to be true as well.

Condition 1   *The preference for symmetry is expressed at or very near birth.* Cell proliferation in the primary visual cortex is maximal at about the time that infants begin to take pleasure from highly symmetric objects. Stimulation of V1 and related downstream visual cortical areas is required for normal brain growth and development, including synaptic pruning (see chapter 8). Laterally symmetric objects represent an ideal form of stimulation during this developmental period, since they have redundant spatial information. Such redundancy may facilitate the increased recognition speed that infants exhibit in processing symmetric relative to asymmetric objects.

Condition 2   *The preference for symmetry generalizes across many object forms.* This condition is critical to the hypothesis that a preference for symmetry originated as a mechanism for ensuring normal brain development and maturation independent of its co-option as a fitness indicator through sexual selection. If a fondness for symmetry in newborns is only fulfilling a role as a potential fitness indicator, we would expect to see the preference emerge strictly in relation to bodies and faces—those objects to which fitness most applies. On the other, if the preference for symmetry in newborns is first and foremost a mechanism to facilitate brain growth, one would expect that all types of symmetric objects are preferred to their asymmetric counterparts. Indeed, the data show the latter to be true, since newborns merely a few months old prefer a wide range of symmetric objects to their asymmetric versions. Such objects include abstract drawings, solid geometric forms, landscape features, faces, and many others.

Condition 3   *Objects that have multiple markers of symmetry, such as faces, are especially pleasurable.* If a general preference for symmetrical features exists independent of the form in which they appear, objects that have multiple salient points that exaggerate this condition should be especially pleasurable to view. Newborns tend to look longer (a proxy for preference) at symmetrically complex objects than at simpler symmetric objects. Newborns also prefer line drawings of faces with the proper symmetry of features preserved more than drawings where the features are shifted and symmetry is broken. Taken together, such evidence indicates that objects with multiple salient features that emphasize symmetry should be even more pleasurable and preferred to simpler, symmetric forms with fewer features.

Condition 4 *Symmetry is a reliable marker of phenotypic quality.* As we have just reviewed, body and facial symmetry are related to improved markers of fertility in both men and women. Both body and facial symmetry have also been shown to be related to better self-reported physical, emotional, and psychological health and well-being. Moreover, independent observers rank people with highly symmetric faces as more conscientious, agreeable, and intelligent, among other positive traits, compared to faces bearing less symmetry. Hence there is evidence that individuals with greater symmetry enjoy better health and fitness and that they are perceived by others as being more likely to possess certain positive personality traits in comparison to their less symmetric counterparts. These and other studies suggest that body and facial symmetry are reliable markers of phenotypic condition, and thus might be used as fitness indicators during mate selection.

Condition 5 *Adults prefer symmetrical bodies and faces.* If body and facial asymmetry are fitness indicators that are used during mate selection, individuals should be able to detect variation in symmetry and prefer it in a potential suitor. Consequently, everything else being equal, individuals with greater body and facial symmetry should be seen as more attractive than those who are less symmetrical. By extension, individuals with the greatest body and facial symmetry (again, all else being equal) should have greater mating success than their asymmetric counterparts.

Condition 6 *Adults prefer symmetrical to asymmetrical versions of most objects in general, even those unrelated to faces and bodies.* If a preference for symmetry emerges in newborns to guide proper brain development through experience-expectant mechanisms, such stimulation need not be in one specific form or another. Faces are excellent forms of stimulation in this scenario since, as we have seen, they possess many cardinal features that activate the pleasure instinct and facilitate neural development. But all symmetric objects should facilitate this process to some degree. Hence, if the preference for symmetry in newborns and children is carried over to adulthood, we would expect it to generalize to other object forms beyond faces and bodies. Let us now turn to a discussion of the evidence for the final two conditions.

Do people have a preference for symmetric mates, and if so, how does this preference impact their selections? In a landmark study, evolutionary psychologist David Buss and his colleagues interviewed more than ten thousand individuals from thirty-seven different cultures about their mating preferences. While men tended to value physical attractiveness slightly more than women in selecting a mate, both sexes from virtually every culture studied ranked appearance as one of the most important factors overall. As we have already seen, there is also remarkable agreement across cultures on aesthetic judgments of facial attractiveness, and even two-month-old infants seem capable of making the distinction. So are people with greater symmetry seen as more attractive?

Perhaps no other research area in evolutionary psychology has received more attention from the media as the study of attractiveness. Several original and replication studies have shown that in general we find potential mates with greater body and facial symmetry to be most sexually attractive. For instance, in comparison to men with high asymmetry, symmetrical men have greater facial attractiveness (as rated by both male and female observers), more sexual partners during a lifetime, and an increased frequency of sexual affairs with partners outside of their primary relationship, begin to have sex earlier in life, and produce disproportionately more copulatory female orgasms in their partners.

In a recent study, biologist Craig Roberts and colleagues from the University of Newcastle found that facial asymmetry and attractiveness correlate positively with each other and with a key marker of immune function. The investigators showed fifty women (aged eighteen to forty-nine years old) color head shots of men with a neutral facial expression. As in similar studies, all photographs were digitally masked so just the face was visible. The women were asked to rank the attractiveness of each face using a seven-point scale. The investigators extracted a composite estimate of fluctuating asymmetry by measuring symmetry at seven distinct bilateral facial landmarks. In an interesting twist, blood samples were also collected from the men and genotyped for heterozygosity at key loci in the major histocompatibility complex (MHC—see chapter 5). Recall from chapter 5 that the MHC genes code for immune cells that identify intruding disease organisms, thus functioning as our immune system's first line of defense. Of importance in this discussion is the fact that MHC genes have upward of a hundred or so different alleles, each providing immunity against different sets of potential disease strains. The level of zygosity refers to how often a specific allele is repeated at different loci within the MHC. It is believed that greater allelic diversity in the MHC leads to a broader resistance to different pathogens. Individuals vary considerably in their degree of heterozygosity, with most people being homozygous at a few alleles. Roberts and his colleagues found that the degree of heterozygosity at key loci in the MHC correlated positively with fluctuating asymmetry in the male faces. Men with greater heterozygosity—who are presumably equipped to fight off a greater variety of invading pathogens than their more homozygous counterparts—had the most symmetrical faces. Moreover, the women rated these faces as the most attractive.

Nonfacial secondary sexual characteristics have also been found to be linked to ratings of attractiveness. As we saw earlier, waist-to-hip ratios in women of approximately .7 are seen by both men and women as most attractive. In other experiments, Devendra Singh demonstrated that breast symmetry is positively correlated with men's judgments of attractiveness as well as their interests in both short- and long-term relationships. Like faces and other secondary sexual characteristics, when observed across a population of individuals, fluctuating asymmetry of breasts tends to be large relative to absolute size (that is, absolute breast size asymmetry divided by breast size). Whereas most body parts exhibit fluctuating asymmetry of no more than 1 percent of the overall body part size, breast asymmetry tends to be closer to 5 percent of absolute size in all cultures that have been studied.

These data indicate that people are able to detect differences in body and facial symmetry, and use this information to guide their choice of potential mates. It is interesting that when asked to define what makes a person attractive, people often say something about a particular look or a particular body part (for example, the eyes). Evidence shows that we use symmetry as an important metric in defining attractiveness and identifying preferred mates, even though we may not consciously recognize this implicit calculation. Interestingly, we use mental calculations of symmetry every day in many other contexts, such as in our appreciation of art, choices of jewelry and clothing, and what consumer products to buy. Let us now examine the broader way the pleasure of symmetry impacts our adult lives.

Symmetry and Aesthetics: An Example of a General Process

Adults are not only drawn to symmetric mates. As would be predicted by the theoretical perspective being discussed throughout this book, the pleasure we take from seeing highly symmetric objects extends beyond bodies and faces. For instance, similar to newborns, adults are able to recognize and process vertically symmetric objects more quickly than objects with similar features that are not symmetric. Moreover, symmetric objects and patterns are preferred by adults over asymmetric versions even if they do not serve any apparent biological function (for example, such as mate selection). Indeed, there is widespread use of symmetric designs for decorative art among cultures diverse in region, ethnicity, and time.

In a series of interesting studies, psychologist Lauren Harris adopted classic abstract designs seen in different cultures (for example, Aonikenk, Navajo, and Yoruba) and manipulated them in terms of their symmetry. In one condition, the geometric shape was manipulated such that two versions of the same form were presented to adult subjects—one perfectly symmetrical, the other asymmetrical. The subjects were asked to "choose the design that is more attractive in each pair of designs." In a second condition, both object shape and coloration were varied—symmetrical in one object and asymmetrical in the comparison. Finally, in the third condition, objects varied in the orientation of their symmetry—vertical versus at forty-five degrees. This third condition was included to replicate previous work that has shown that both newborns and adults recognize objects with vertical symmetry more quickly than other orientations.

Harris and her colleagues found that in all comparisons the symmetric version was seen as being more attractive than the asymmetric counterpart. This suggests that symmetry is preferred in nonbiological signals outside the context of fitness, and that the preference is robust since it occurred with respect to the primary features of shape, color, and orientation. Condition two, in which both color and shape symmetry were varied, was found to have the largest effect size (differences in the mean number of designs chosen as most attractive in each condition) of all the conditions. This finding is exactly what would be predicted from the theory being considered, namely that there are several core preferences (across all sensory domains) that emerge during development that facilitate normal brain growth and maturation. The pleasure instinct prods us to seek these basic stimulus forms to fine-tune each sensory system to the environment in which the individual resides. These core features are additive in the sense that objects with multiple pleasure-inducing stimulus forms should be preferred to those objects with fewer forms. Hence it is more pleasurable to look at a real face (in both newborns and adults) with smooth skin, high contrast and concentricity, and symmetry than a simple line drawing of a face with just symmetry. In this case we see that color symmetry acts additively with shape symmetry to produce an even greater preference than either would have alone.

Interestingly, Harris and her colleagues also examined the impact of symmetric and asymmetric facial painting on overall facial attractiveness, noting that such practices are common across geographically diverse tribal societies (for example, the Selk'nam of South America, the Huli of Papua New Guinea, the Kikuyu of Africa, and Blackfoot Indians of North America). Again, they had three conditions. In the first condition, unpainted symmetric versions of a face were compared against unpainted asymmetric versions of the same face. In the second condition, symmetric faces with laterally symmetrical paint designs were compared against asymmetric faces with laterally asymmetric designs. In the final condition, symmetrical faces with asymmetrical paint were compared to asymmetric faces with symmetric paint. Subjects were asked to "choose the face that is physically more attractive in each pair of faces."

The researchers found that symmetric faces with symmetric paint were viewed as the most attractive in terms of absolute preference across all of the conditions. They also found, as expected, that unpainted symmetric faces were preferred to unpainted asymmetric faces. Interestingly, the application of an asymmetric design to a symmetric face decreased its attractiveness, while the application of a symmetric design increased the attractiveness of asymmetric faces. This pattern of results indicates that additional symmetry features that have no association with overall fitness or phenotypic quality have an additive effect on preference ratings of biologically relevant features that do indicate fitness.

Our innate preferences for symmetry and proportion that impact everyday behaviors are but two simple spatial examples of what I believe represent a general process. As mentioned in earlier chapters, the pleasure instinct creates strong preferences for distinct stimulus features in every sensory domain. The developing brain is thirsty for particular experiences that optimize both synaptogenesis and synaptic pruning. Some feature preferences crafted by the pleasure instinct are likely carried into adulthood unaltered and continue to steer our everyday behaviors and choices in subtle and not so subtle ways (for example, our love of sugars and fats). But my sense is that many of the preferences that facilitate brain development (arguably adaptations driven by natural selection) have also been amplified at some point in our phylogenetic history by sexual selection processes. In the next chapter we will discuss a temporal example and examine the complicated manner in which our preference for repetition and rhythm is expressed in our everyday lives.

Chapter 10

Pleasure from Repetition and Rhythm

It appears probable that the progenitors of man,
either males or females or both sexes, before acquiring
the power of expressing their mutual love in
articulate language, endeavored to charm each other
with musical notes and rhythm.

—Charles Darwin, *The Descent of Man*

Existence equals pulsation.

—Ellen Dissanayake, *Homoaestheticus*

As we saw earlier, our innate preferences for proportion and symmetry have been crafted by the pleasure instinct as a means to encourage newborns, toddlers, and older children to seek out optimal forms of spatial stimulation to fine-tune the developing visual system during synaptogenesis and synaptic pruning (see chapter 8). In the previous chapter, we discussed how this preference has been co-opted during our species' phylogenetic history through sexual selection, since such visual features can also be used as fitness indicators for mate identification and mate choice. In this chapter we will consider a temporal example and examine how our innate preference for repetition and rhythm impact many of our everyday behaviors, some of which may also be magnified through sexual selection.

The development of the primary auditory cortex and associated brain structures of all mammals depends critically on the precisely timed expression of key genes triggered when the organism experiences environmentally relevant stimuli to fine-tune the system. As was remarked earlier, the details of development are not in the genes, but rather in the patterns of gene expression. Between the twenty-fifth and thirtieth weeks of gestation, fetuses become sensitive to sounds and will move in relation to Mother's voice especially. But beyond Mother's voice, they also hear the steady rhythm of her heartbeat and respiration.

After birth, infants continue to seek out certain forms of auditory stimulation, particularly those that are repetitious and rhythmic. For instance, newborns and infants are extremely sensitive to the prosodic elements of speech, those that are rich in emotional meaning. Of course, prosody is the backbeat of the well-known singsong style of motherese that dominates parent-infant dialogue during the first year of life, with its emphasis on simple pitch contours, broad pitch range, and syllable repetition. Experiencing acoustic rhythm during the latter period of gestation and then as a newborn and toddler seems to be a key requirement for normal growth and maturation of the neural systems that process auditory information. Recall from chapter 7 the experiments by neurobiologists Michael Merzenich and Edward Chang of the University of California at San Francisco showing that newborn rats failed to develop a normal auditory cortex when reared in an environment that consisted of continuous white noise. After only a few months, the scientists found significant physiological and anatomical abnormalities in the auditory cortex of the noise-reared rats when compared to rats raised in a normal acoustic environment. Moreover, these abnormalities persisted long after the experiment ended. Most interesting of all, when the noise-reared rats were later exposed to repetitious and highly structured sounds, their auditory cortex rewired and they regained most of the structural and physiological markers that were observed in normal rats.

Rhythm also seems to have a calming effect on newborns and children through several sensory modes. For instance, it is commonly known that older infants, children, and even adults employ various "comfort actions" of rocking back and forth, or doing some other form of rhythmic, repetitive action to relieve tension or stress. Various religious groups practice meditation and prayer using a combination of rhythmic chanting, rocking, and repetition of key phrases. Babies as young as four months old are calmed by repetitive sounds containing consonant rather than dissonant intervals. Auditory rhythms built using intervals such as the "perfect fifth," with a pitch difference of seven semitones, or the "perfect fourth," with a pitch difference of five semitones, are calming to babies and adults from all cultures. Hence the experience of repetition and rhythm is critically important during brain development and can have a marked impact on the behavior of babies, children, and adults.

Indeed, most parents might observe that their babies tend to self-stimulate their brain growth by producing highly repetitious and rhythmic sounds. Ren, our latest arrival who is now seven months old, does exactly the things Kai did at this age. She seems to take great pleasure in quickly repeating the same syllable, often with slight variation in pitch toward the end of a series—her own, babbling version of motherese. She also demonstrates innovation by abruptly changing the melody slightly by alternating syllables with different pitch—beep, bop, beep, bop. She has not experienced all of these individual melodies, yet she is clearly able to take basic sounds and string them into novel series, and she does so with great joy.

An early preference for experiencing repetitious and rhythmic stimulation could have a profound effect on many forms of adult behavior. Indeed, such preferences may lead to the development of our fondness for a host of pleasurable phenomena such as music production and perception, dance, poetry, and ritualistic behaviors, to name but a few. As an example, let us consider music, since it has received the most recent attention in terms of its evolutionary origins.

The notion of the extended phenotype was made popular by the evolutionary biologist Richard Dawkins. In general, the phenotype refers to an observable quality of an organism such as its development, morphology, or behavior. This is distinguished from the term genotype, which refers to the set of hereditary instructions an organism contains. It is fairly well accepted that the phenotype can be influenced by three primary factors—the organism's genotype, transmitted epigenetic factors, and nonhereditary environmental factors. Dawkins argued that the classic idea of the phenotype was too restrictive, since it focused primarily on the phenotypic expression of genes in the organism's own body. In his popular book *The Extended Phenotype*, he wrote, "An animal's behaviour tends to maximize the survival of the genes for that behaviour, whether or not those genes happen to be in the body of the particular animal performing it." Genes have a long reach in that they may code preferences for a trait or behavior in one organism that, through sexual selection, promotes the emergence of the trait or behavior in others. This is a similar scenario to what we saw with a preference for symmetry and may help us understand how an innate preference for acoustic repetition and rhythm, in the service of brain development, could lead to the evolution of music production.

An appetite for music is ubiquitous among humans. It seems that evidence of music production pops up wherever there are relatively stable human social groups. And music production has been around for quite a while. Percussive and flutelike instruments have been found at *Homo sapiens* sites throughout Europe and Asia dating as far back as a hundred thousand years. Archaeologists have also found a bone flute at a Neanderthal site near Idrija, in northwestern Slovenia. This flute was made from the polished thighbone of a bear and consisted of four carefully aligned holes drilled into one side. Strikingly similar bone flutes have been discovered at *Homo sapiens* sites dated between forty thousand and eighty thousand years old.

Traditionally, evolutionary biologists have limited their forays into the study of how music may have evolved. Musicologists and theorists, similarly, have not been greatly concerned with the origins of music or its adaptive value. My sense is that this neglect from both sides is primarily because music has not been viewed as a behavior that has obvious survival value. But as we have discussed in previous chapters, evolution is driven by genes that survive long enough in one organism to be passed on to another organism ad infinitum. Mere survival in a single host is not the endgame of evolution. Darwin himself believed that the evolution of music production and enjoyment was best understood as a sexually selected adaptation.

In his recent book *The Singing Neanderthals*, anthropologist Steven Mithen argued that the conventional wisdom that music has no direct survival value is dead wrong. Rather, he suggests that in addition to its potential importance in sexual selection, our prelinguistic ancestors relied on music as a means to facilitate communication and cooperation. You might think that the term "cooperation" suggests a need for employing group selection as a theoretical driver of music as an adaptation.* On the contrary, there is no need to invoke group selection, since one could argue that music production and perception have direct benefits for individuals (and the propagation of their genes) with a strong sense of alliance to the group. Such individuals are better protected from predators and enjoy the many survival advantages tied to group participation such as coordinated foraging, technology sharing, and rearing of offspring.

The evolutionary psychologist Geoffrey Miller of the University of New Mexico has taken a slightly different route. Following Darwin's lead, he suggests that music production and perception resulted strictly from sexual selection mechanisms. In an interesting retrospective study, Miller took random samples from entries in major musical encyclopedias (more than eighteen hundred samples of jazz albums,

---

*Recall from our earlier discussions that group selection is a mechanism usually dismissed by evolutionary biologists, since it requires rather extreme and unlikely conditions to operate in the real world.

more than fifteen hundred rock albums, and more than thirty-eight hundred classic music works). He notes that from this sample, "males produced about 10 times as much music as females, and their musical output peaked in young adulthood, around age 30, near the time of peak mating effort and peak mating activity." This pattern of results is consistent with a large body of data showing that adaptations for courtship display (across many different species) tend to be sexually dimorphic (exaggerated in one sex) and emerge when the organism reaches sexual maturity. Of course, this study has limitations in that only those individuals with a high degree of professional success as musicians have been recorded in such encyclopedias. There may be key differences between the demographic distributions of everyday, generic musicians who just like to play and listen to music for the fun of fit and the distributions of professionals who have achieved some degree of popular success.

Music production on the African savanna during the time of our hunter-gatherer ancestors was certainly not the complicated, techno- logical affair it is today. Most likely it was a group activity, as it is in many modern tribal societies. Moreover, if ancestral music is similar to what one sees in modern hunter-gatherer societies, it was prob- ably typically accompanied by dance. It is difficult to imagine the hunter-gatherer equivalent to the modern-day jazz concert, with the performers up front surrounded by a much larger group of pas- sive listeners just sitting around. The suggestion I favor is that music and dance were closely aligned in ancient times and served as proxies for fitness, just like symmetry. The primary difference is that body symmetry is a morphological phenotypic trait that can be directly correlated to a host of fitness metrics (see chapter 9), whereas music production and dance are phenotypic behaviors. Thus it might prove more difficult to show how music production is related to fitness. But as Dawkins and others have observed, examples abound in the animal signaling literature where a call, song, howl, or growl has evolved as a courtship display through sexual selection.

Might music production and perception be viewed as having evolved through sexual selection processes as a courtship display? Let us suppose that a preference for repetitious and rhythmic acoustic sounds first emerged as a survival mechanism linked to advantages in brain development. As we have seen in earlier chapters, the preference for repetitious and rhythmic sounds by newborns and babies prompts them to experience as much of these stimuli as possible, which in turn facilitates normal brain growth and maturation during synaptogenesis and synaptic pruning. Studies in rodents and primates have found that a lack of exposure to patterned acoustic stimuli immediately after birth and the months that follow has profound effects on the physiological and anatomical brain structures responsible for audition. These changes to normal brain physiology and anatomy produce marked deficits in normal hearing and functioning. Organisms with deficits in normal audition might have a greater risk of predation (since they can't hear a predator charging from behind) and other general safety concerns that decrease their likelihood of surviving to reproductive age. A receiver bias such as a preference for repetitious and rhythmic sounds would thus have survival value and could certainly have evolved through natural selection.

But how does the pleasure we naturally find in rhythm lead to the eventual production of Schubert's *String Quintet in C Major* or *Rhapsody in Blue*? There is evidence suggesting that the evolution of music production has been driven by both natural and sexual selection mechanisms. Clearly, music has effects on social communication and cohesiveness that may benefit the individual's likelihood of survival. My sense, however, is that sexual selection has played as important a role in the evolution of music production as natural selection. Just as was the case for body symmetry, it is possible that an innate receiver bias in the form of a preference for repetitious and rhythmic sounds could have been co-opted as a courtship display through sexual selection. The fact that music is such an elaborate and complex adaptation makes for an even more compelling case. If a preference for rhythmic sounds becomes genetically correlated with the production of rhythmic sounds as a behavioral trait, a positive feedback loop could occur, leading to a Fisherian runaway process. Comparable to the peacock's plume, this might lead to ever more elaborate displays to vie for the attention of the opposite sex. The trouble with this argument is that it is considerably easier to suggest that music evolved as a sexually selected courtship display than it is to find actual data supporting the claim.

Ironically, going back all the way to Darwin, there are far better examples of how animal signaling—such as calls and songs—serve as courtship displays in species of frogs and birds than how making music may serve a similar purpose in humans. Many species of birds, whales, and primates (for example, gibbons) use song as a sexual display during breeding season. Some songbirds such as the winter wren use bits and pieces from songs they have heard, and reassemble them to form new phrases. That is, they exhibit learning similar to that seen in human music making. Hence, winter wrens can often sing hundreds of different songs. There is now evidence that females from several bird species such as blackbirds, mockingbirds, and warblers prefer males that generate larger song repertoires. Many species of birds also sing songs that have key components such as refrains, symmetry, and reprises similar to what is seen in human music.

So there is a precedent for using signals that have some components in common with human music as a form of courtship display during mating. To demonstrate that human musicality evolved through sexual selection as a result of an initial receiver bias for repetition and rhythmicity, however, we need strong evidence to support a number of basic necessary conditions (similar to the conditions we generated when discussing symmetry and proportion).

Condition 1   *The preference for repetition and rhythmicity is expressed at or very near birth.* We have provided key evidence for this condition existing in audition in chapter 7 and for other sensory modalities in earlier chapters.

Condition 2   *The preference for repetition and rhythmicity generalizes across many object forms.* As we saw in earlier chapters, infants take pleasure in both perceiving and producing repetition and rhythm in a number of different domains, including vision, somatic, vestibular (for example, rocking back and forth), and audition. Hence this condition is likely met given the accumulated evidence.

Condition 3   *Music has signal properties (qualitative and quantitative) that are similar to the repetitious and rhythmic stimuli that infants enjoy and produce.* One way of providing evidence for this condition is to examine the statistical properties of the rhythmic auditory signals most enjoyed by infants and demonstrate a correspondence in music. There is some evidence for this already. As we have seen, infants and adults alike prefer rhythms built from consonant intervals such as the perfect fifth and perfect fourth, with small pitch differences in neighboring tones over dissonant intervals with high pitch differences. It has also been shown that infants prefer rhythms with sharply rising or falling pitch contours. This signal quality is common to both language (that is, prosodic cues) and music. Other evidence might come from showing that the repetitious and rhythmic auditory signals that infants generate as a form of self-stimulation have similar properties to those seen in music. To date this type of study has been limited to demonstrating that some of the key components of musical expression also occur in infant babbling—for instance, repetition, abrupt changes in pitch contour, reprises, and refrains.

Condition 4   *Music production is a reliable marker of phenotypic quality.* Since music and dance have costs associated with their production, this behavior might serve as a potential fitness indicator. Fitness itself is a vague term. Most often when we hear the word "fitness," we think of proxies such as general health, avoiding certain illnesses, heterozygosity in select genes, or any number of other traits. But for *Homo sapiens* with sophisticated cognitive functioning, fitness of the mind may also have been an important indicator. As we have seen, handicaps are often reliable fitness indicators because they are generated at the expense of diverting precious energy from basic system functioning (for example, from growth and immunocompetence) and are often very conspicuous signals that could attract the attention of predators (they are called ornaments for good reason). Music production and dance require large reserves of metabolic energy even for short periods, but we know that modern tribal societies often have ritualistic music and dance that can last many hours and even days. The closest Western equivalent is probably the all-night rave, where young adults typically in their twenties dance nearly continuously from dusk to dawn.

Music and dance production also consume vast amounts of metabolic energy used by the brain to derive creative expressions that will attract the attention of potential mates. The overwhelming majority of energetic costs associated with being a big-brained hominid are linked to that big brain. No body part or system even comes close in its energy requirements to the ongoing metabolic demands of our brain. Hence traits that are reliably linked to increasing the already high energy demands of the brain would be honest indicators of fitness in this broader sense.

Miller has made some suggestions in this regard. "Dancing reveals aerobic fitness, coordination, strength, and health. Because nervousness interferes with fine motor control, including voice control, singing in key may reveal self-confidence, status, and extroversion. Rhythm may reveal the brain's capacity for sequencing complex movements reliably, and the efficiency and flexibility of the brain's 'central pattern generators.' Likewise, virtuosic performance of instrumental music may reveal motor coordination, capacity for automating complex learned behaviors, and having the time to practice." If some of these putative associations or others could be demonstrated through empirical studies, this would represent a major step forward in showing that the evolution of music was indeed driven, at least in part, through sexual selection.

Condition 5   Adults who can produce music enjoy greater mating success (when all other potential confounding variables are controlled) than their unmusical counterparts. As with condition 4, this condition might also be extended to include dance.

There is, of course, the anecdotal evidence about rock stars having magnetic-like attraction for members of the opposite sex, but I suspect that there are so many confounding variables (for example, confidence, extroversion, wealth, exposure, fame) that it is practically impossible to draw any meaningful conclusions from such stories. To demonstrate that this condition is true, we need evidence from prospective, well-controlled experimental studies. One line of experimentation should be designed to determine if individuals who exhibit the best dance or music production ability are viewed as more attractive than their counterparts after being matched for certain covariables such as age, gender, and key personality traits. The second line of experimentation would support condition 4 as well as condition 5 by testing whether the individuals who are seen as being the most attractive are also seen as being the fittest. Such experiments should also attempt to answer the specific dimensions of fitness that are particularly relevant to music production and dance. To my knowledge, such studies have not been conducted to date and await a motivated researcher.

Condition 6 *Adults prefer repetition and rhythmic stimulation in many forms, even those unrelated to music production and perception.* A growing body of data indicates that adults are attracted to repetition and rhythm in many contexts across many sensory domains. Children and adults take pleasure from certain comfort actions such as rocking back and forth or producing repetitive movements of one body part or another. Such movements are pleasing in their own right but also reduce tension and stress. Dancing, running, walking, swimming, sex, and a host of related activities are other forms of repetitive motor behavior that clearly please adults.

Language production and associated social dynamics have a very specific rhythm that can be pleasurable. Indeed, experiments have demonstrated that when speaking, people tend to choose words that fit rhythmically into their statement—a type of melodic intonation of spoken language. The pleasure some individuals find in poetry, which depends strongly on rhythm and meter, is another example of our attraction to repetition and rhythmicity.

Many nonmusical sounds of nature that are rhythmical are pleasurable to adults and children. Judging by their impressive sales, quite a few people fall asleep each night to the soothing beat of the ocean produced by a sound generator. Other options include the sound of crickets, rhythmic winds, flowing brooks, and birds. As long as there is sufficient variation in the sequence and the sounds are rich enough to reflect the natural world, the rhythm is very pleasing.

Although many studies have examined our proclivity to prefer temporal order to chaos, clearly there is a need for more systematic research to map out the full extent of the sensory domains involved. I suspect that we have just touched the surface in really understanding how this preference presents itself in everyday behaviors.

Hence we have fairly good evidence supporting conditions 1, 2, 3, and 6, and we need quite a bit of additional data to convincingly support conditions 4 and 5. But I am convinced that evidence will emerge if we make the effort to conduct carefully controlled studies.

To this point, we have discussed two very different examples—one spatial and one temporal—that illustrate the way the pleasure instinct can impact our everyday lives and behaviors. In the next chapter we will consider the manner in which the pleasure instinct places high costs on those individuals who abuse it. We are all equipped with brains that have evolved to face specific challenges and circumstances from our ancestral past. Many of these challenges and the conditions in which they originated are quite different from, and in some cases in direct opposition to, those that exist in the modern world. In a real sense, we are all of another time. The innate preferences that have been forged by the pleasure instinct to help facilitate brain growth and maturation have consequences far beyond our love of symmetry, proportion, rhythm, and repetition (to name just a few). Let us now turn toward the darker side of the pleasure instinct—addiction.

Chapter 11

Homo Addictus

*Every form of addiction is bad, no matter whether
the narcotic be alcohol or morphine or idealism.*

—Carl Jung, 1963

Vices are sometimes only virtues carried to excess!

—Charles Dickens, 1848

Most people have no idea how much their brain changes on a daily basis. As you read these words, distinct neural ensembles are communicating with one another, shuttling electrical impulses across brain space. In the process some of these neural paths become strengthened and others are weakened. This collective pattern of brain activity creates a map or neural representation of the information being learned. As we have seen in previous chapters, some things are generally easy to learn if they are related to an organism's overall fitness or survival. Information not directly related to important selection factors may be more difficult to learn if it has little or no fitness relevance. The degree of difficulty in learning something is generally measured by seeing how long it takes to master the new information. For instance, if you become nauseous after eating dinner at a particular restaurant, you do not need additional meals to form the association between sickness and the local greasy pit. This is true for all mammals. Rats that are made sick by ingesting tainted food will avoid the food and location where it was consumed after a single experience. In contrast, it takes much longer to learn and remember multiplication tables or word definitions, information that—one might argue—is not directly relevant to survival or reproductive success.

As a young professor, my scientific interests focused on under-standing the changes that occur in the brain as something is learned and remembered. Deep in the medial portion of your temporal lobe, there is an area called the hippocampal formation, which lights up like Carnivale as you learn new information and begin to store it into long-term memory. A great deal is now understood about the cellular and biochemical changes that occur in the hippocampus and related structures during learning and memory. Changes of this sort are generally referred to as neural plasticity, a phenomenon associated with a host of normal and abnormal conditions.

Many scientists who study neural plasticity also study addiction, since it is believed that the transition from casual substance use to dependency is accompanied by distinct changes in the way disparate brain regions communicate with one another. A number of modern treatments for addiction, as we'll see, focus on blocking these changes in neural communication. Such phenomena can be studied readily in mice and rats, although there are obvious limitations in making the conceptual leap from animal models of addiction to understanding the disease process in humans. My approach to help bridge this gap was to volunteer at a local adolescent facility for substance abuse to hear about the addiction process from people who have experienced it firsthand.

The building in which I was eventually to spend so many after-noons was an old converted Victorian house on the outskirts of downtown. I learned quickly that the treatment model at this facil-ity was holistic. Kids aged twelve to seventeen years resided in the house for therapeutic periods ranging from roughly three to twelve months. A typical day included meals, four hours of school, individual and group therapy, medical appointments with physicians and psychiatrists, meetings with legal counselors if required, and family visits. Kids came from all over the West and for a variety of reasons. Some had been in trouble with gangs and been arrested repeatedly. Others were at the house for behavioral problems at school or home. A common theme among the kids was substance abuse that could involve alcohol and/or controlled substances, including prescription medicines.

From the beginning, I was deeply moved by the emotional stories I heard from the residents. Several common topics came up again and again, including childhood traumas such as physical, sexual, and verbal abuse. Other kids were impacted severely by a single early event such as the death of a parent or sibling. After several months I began to see patterns in an individual's choice of drugs that seemed to map onto the particular circumstances that surrounded his or her life.

Alberto was a seventeen-year-old boy who had been repeatedly plucked off the streets of Phoenix by authorities for crimes related to gang activity. When I first met him, he didn't seem violent, but I knew Alberto had been arrested at least once for assault on a rival gang member. He wasn't a terribly big guy and, to me, he seemed almost easygoing. If anything, he projected a sense of detachment bordering on apathy.

Each resident participated in group sessions three times a week. A session typically began with each resident giving a brief update on his or her current state and bringing up any problems to the group. Alberto never seemed to have any problems. Like many new kids, he seemed to think of group therapy as a chore that was best done as quickly as possible or avoided entirely. After the update period, the group would focus on one person and explore the circumstances that brought them to the house. During his first turn Alberto seemed painfully uncomfortable. He appeared unable to focus and became more and more frustrated with each passing minute. The group, however, had seen this before and gave him time. Gradually he began to tell his story.

Alberto came to the United States from Mexico when he was eight years old. He and his mother moved into a small, two-bedroom apartment with other family members, including his aunt and uncle and their four children. He described his uncle as a chronic alcoholic with a quick temper who physically abused him and his cousins fairly regularly. Alberto attended school for a couple of years when he first immigrated to the States, but dropped out and got more involved in gang life in his early teens. By the time he was thirteen years old, Alberto had tried almost every drug available on the street and was selling methamphetamine with a crew of other kids and a connection out of Los Angeles that could be traced back to Mexico. His favorite drugs were methamphetamine and cocaine, both of which he consumed regularly.

One summer night, after a day of meth binging, he had a psychotic episode. He described the experience as a waking dream in which he heard angry voices yelling at him, but he could not understand exactly what was being said. He also felt worms crawling under his skin, and he picked violently at his arms, neck, and face until they bled. At some point in the night Alberto had a grand mal seizure and was raced to a local emergency room. The ER visit was followed by police custody. After several similar experiences, arrests, and detoxifications, Alberto was sent to our little house for full-time residential care.

The withdrawal state that he felt was fairly typical of cocaine and methamphetamine use: low arousal and a general sense of malaise. Almost all methamphetamine or cocaine users appear lethargic and extremely apathetic following detoxification. In contrast, Alberto described the feeling of a meth-induced high as being like a bull—strong enough to take on anything or anybody. It also gave him enough energy to keep him awake for days on end. The best part for him was often the anticipated high and then the feeling that nothing could go wrong once the drug took effect. In the year that I worked at the house, I saw many ex-gang members. Almost all of them were addicted to methamphetamine and described a sense of invincibility while on the drug that made it particularly attractive given the toughness of gang life.

Although the most common addiction in the house was to meth-amphetamine, there were also a number of residents addicted to heroin or morphine. Those addicted to heroin or morphine often had noticeably different life circumstances surrounding their drug use compared to those using methamphetamine.

Christine was a petite blonde who could easily be mistaken for the class valedictorian. She was often described by her peers as "bubbly," instantly likable, and very smart. She came to the house from Las Vegas after running from three other rehabilitation programs. Her guardian hoped that bringing her out of state away from friends to a residential program might prove more effective in addressing her heroin addiction. I first met Christine in a group therapy session. Based on appear-ance alone, most people would have never guessed that she was a heroin addict. Nor would they likely be able to fathom the strange world in which she was immersed while using the drug.

Contrary to many of the kids at the house, Christine actually embraced the program and was eager to participate. In group ses-sions we began to learn about her surprising past. She was born just outside San Francisco, but moved with relatives to Las Vegas after her parents were killed in an automobile accident. In Vegas Christine often felt like an interloper, living with her grandmother and ail-ing grandfather. Shortly after arriving, her grandfather died and her grandmother sank into a deep depression. Christine was thirteen when her grandmother committed suicide, leaving her to fend for herself. She dropped out of school and lived on the street with a small group of other homeless teenagers. Her new life consisted of prostitution and just trying to stay alive. One day a friend showed up with several small vials of pure morphine stolen from a local hospital and asked if she'd like to join her. Christine had tried other drugs by then, including pot, methamphetamine, and a host of pre-scription drugs. She described her first morphine use as a turning point in her life. She had never had a high like this before and felt an instantaneous warmth come over her entire body—almost as if a security blanket was being tucked around her by her long-lost par-ents. She felt safe and, for the first time in as far back as she could recollect, less anxious and sad about her life. Before morphine, she constantly worried about everything; now all that was gone.

Christine quickly made the jump from morphine to heroin and started to get involved in petty theft, mostly stealing jewelry and wallets from hotel rooms on the less glamorous side of town. After her second arrest she was sent to a juvenile detention program that was followed by her first rehabilitation program. She was arrested a third time for prostitution less than three weeks after completing the initial rehab.

In group sessions, Christine described her gravitation toward heroin use as a logical choice, almost as if she were a pharmacist matching a treatment to a particular ailment. Her problem, of course, was extreme anxiety. The typical uppers such as speed, methamphetamine, and cocaine always seemed to worsen this state. Christine learned through her own trial and error that morphine, heroin, and sex were all ways to ameliorate this anxiety and unrest. This process was not altogether different from the experiences of Alberto, who learned that methamphetamine often made him feel more confident and brave in gang-related circumstances that can easily be described as perilous. Time and again I heard similar descriptions of how a resident came to focus on a particular drug or combination. It was not long before my understanding of addiction at the neural level started to align with what I was hearing from these kids, who had lived the experiences. The stories I heard mapped well onto theories about how different brain structures sensitive to addictive substances modulate the pleasure instinct.

At present, there are at least three major theories of addiction, each involving biological and psychological components. We will discuss these in a bit, but before we do, it may be instructive to first think about addiction as a process that interacts with emotional systems— both biological and psychological in nature.

Researchers have often found it useful to separate emotions into two basic processes, one that represents the valence of the state (positive or negative) and another that describes the level of physical arousal (high arousal or low arousal). In this two-dimensional model, one can have positive feelings involving high arousal. This state occurs when the arrival of some positive event (for example, a loved one or the smell of a tasty cheeseburger) triggers pleasurable feelings. Contrasting this, the arrival of a negative event (for example, bad news or the immediate threat of physical harm) can induce a feeling of dread or anxiety.

It's important to remember that in this model, pleasurable feelings can also be elicited in the low-arousal state by the removal of a previous threat. Likewise, the removal or loss of a potentially useful event can lead to negative emotions. Psychologists like this model because it aligns well with experimental findings. For instance, the pleasure associated with the introduction of a positive stimulus is typically accompanied by relative increases in clinical indicators of arousal (for example, blood pressure and cortisol levels). Likewise, pleasure elicited by the removal of a constant threat (or negative stimulus) is usually associated with relative decreases in these same clinical indicators. In each case we have the same emotional end point, but an asymmetry in how one arrives at the destination.

Neuroscientists who study emotions from an evolutionary perspective also like this two-dimensional model since it corresponds well with the idea that emotions are important for identifying fitness indicators in one's environment. An example of this is the high positive correlation between an individual's facial symmetry and his or her perceived attractiveness by others (see chapter 9). Our world is full of fitness indicators that range from those that can be used to determine the ripeness of fruit to others that allow us to choose a suitable mate. This simple two-dimensional model of emotions extends naturally to the traditional view that hedonic states evolved as internal measurement devices for assessing fitness (see chapter 9). In this view, a given stimulus carries emotional value only if it can serve (indirectly or directly) as a fitness indicator. You might argue that this is a drastically oversimplified view of emotions, and I would agree. We will use it here only to introduce a perspective for understanding how the pleasure instinct relates to the initial attraction and subsequent abuse of drugs and other potentially addictive phenomena.

In this two-dimensional fitness model, pleasurable feelings occur with the presence of fitness benefits or with the absence of fitness decrements. Negative feelings occur with the presence of fitness decrements or with the absence of fitness increments. This model, although simple, is consistent with a large body of experimental findings in humans, nonhuman primates, and mammals.

We all know, however, that a fitness indicator that is useful at one point in historical time may not be useful at another time if the environment in which natural selection takes place changes dramatically. For example, in earlier chapters we learned that our intrinsic fondness for sweets in the forms of fructose and lactose serves an important function of feeding the metabolic machinery of each cell in our body. This fondness, which was forged by selection pressures in our prehistoric hunter-gatherer days, has turned into a pathological condition in modern environments with the advent of refined sugars. It now contributes to a number of modern health problems, including obesity, diabetes, and heart disease, to name just a few.

While this particular example makes sense, one might ask how this general process extends to a fondness for drugs or alcohol. Lactose and fructose were clearly available in our hunter-gatherer days, so it was at least possible that they might be used as selection factors. Those who could identify and consume these resources stood a better chance at surviving to reproductive age. But were alcohol and other psychoactive compounds available? Certainly we can't expect the synthetic forms we have today to have existed during hunter-gatherer times, but what about their precursors? If such substances did not exist during ancestral times, they never could have been used as selection factors, and hence an evolutionary theory of addiction based on the hedonic model would not make much sense.

People often think psychoactive drugs are modern phenomena; they are not. They are a modern problem, to be sure, but their precursors have coevolved with hominoids through the millennia. Many anthropologists have pointed out that *Homo sapiens* have enjoyed a coevolutionary relationship with psychotropic plants for millions of years. In this coevolutionary arms race, mammals have evolved mechanisms to metabolize certain plant substances, and at the same time, plants have evolved toxins that mimic the chemical structure of many endogenous neurotransmitters and neuropeptides. For instance, *Areca catechu*, commonly known as betel nut, was being used at least thirteen thousand to fifteen thousand years ago in ancient Timor. You have probably never heard of betel nut, but it is currently the fourth most commonly used drug on the planet following nicotine, ethanol, and caffeine. There is also evidence that nicotine was being extracted from pituri plants by indigenous Australians in Queensland some forty thousand years ago.

An open question concerns how and when these substances were used. For many psychoactive substances, there is archaeological evidence that they were used in ceremonial contexts, but there is also evidence that they were simply everyday food sources and used for medicinal purposes. Our relationship with alcohol probably goes back even farther than the drugs just mentioned. Virtually every species that ingests fermenting fruit is subject to low levels of ethanol exposure. Indeed, the anthropoid diet has been predominantly frugivorous (fruit-eating) for some forty million years, suggesting that ethanol exposure is old and prevalent in our prehistory. Temperate-zone fruit sources have been shown to manifest ethanol concentrations ranging from 0 to 12 percent. Comparative studies have found that as most temperate fruit ripen, both their ethanol and natural sugar contents increase. Consequently, mammals that were consistently able to identify and consume fruits enjoyed the fitness benefits of fructose, but also ingested low levels of ethanol as part of their diet. Some anthropologists have suggested that ethanol plumes may have even been used by early mammals to identify fermenting fruit, making the identification of environmental ethanol a fitness indicator. Regardless of their use, hominids have clearly had a long relationship with plant-derived psychoactive compounds.

At this point we know three very important things. First, psychoactive substances were likely consumed quite commonly in ancestral environments. Second, mammals have evolved distinct mechanisms for detecting and consuming these substances. Third, these substances may have had fitness value in the form of medicines, food supplements, or a means to find either of the two. Hence our simple two-dimensional hedonic fitness model of emotions may apply to these psychoactive compounds. So how do emotions play a role in addiction? In particular, why does the pleasure instinct nudge some of us toward addiction, but not all of us?

The Many Faces of Vice

It is quite popular to experiment with potentially addictive drugs. More than 60 percent of Americans have tried an illicit substance at least once in their lifetime, and if alcohol is included, the number rises to more than 90 percent. But, of course, only a very small percentage of people who actually try a potentially addictive drug become addicted. For instance, recent studies have found that even for a highly addictive drug such as cocaine, only 15 percent of users become addicted within the first ten years of use. The addiction literature often focuses so intently on the particular psychological and biological mechanisms that might be responsible for the addictive process that we seldom ask the simple question as to why so relatively few users ever become addicted. An evolutionary perspective may prove particularly useful in addressing this question, since it assumes that we are *all* susceptible to addiction, not just some of us. Another question that can be addressed from an evolutionary view concerns why there are so many different forms of addiction. Are there psychological or biological mechanisms that are common roots for all forms of addiction, whether the compulsion is to use heroin, eat fried food, or gamble?

What are the different kinds of addiction? The answer changes depending on whom you ask. Certainly there are the classics that we all typically think of when we talk about chemical addictions such as drugs and alcohol. But what about other activities, such as food, sex, video games, surfing the Internet, thrill-seeking, shopping, and so forth, that may share common points with the more traditional forms? Let us look at the leading theories of what addiction is and how it forms. This will provide greater context for understanding how the pleasure instinct may contribute to the casual use of certain substances and how this use may transition to full-blown addiction.

There is an enormous literature in this area, but three major theories have stood the test of time. Each tries to explain the psychological variables and processes that govern the transition from casual to compulsive substance use. They are: (1) the *classic hedonic view* that drugs are taken for the pleasure they provide the user and that unpleasant withdrawal symptoms are the primary cause of addiction; (2) the *aberrant learning perspective*, which holds that addiction results from the formation of pathological stimulus-response associations; and (3) the *loss of inhibitory control theory*, which suggests that the brain systems that usually regulate impulsivity may be impaired, resulting in greater susceptibility to substances that provide immediate gratification. I will introduce and contrast these theories with a fourth, the *modified or modern hedonic view*, based on recent findings that the neural systems responsible for "wanting" a drug are different from the systems that control "liking" a drug.

The classic hedonic explanation for addiction dates to the 1940s, but it wasn't until the work of Richard Solomon and his colleagues in the 1970s that formal theories were first developed and tested. The basic idea is that we take drugs because they bring us pleasure. Repeated exposure to the same drug, however, leads to tolerance such that ever-increasing doses are needed to get the same high. The same homeostatic neural mechanisms that lead to tolerance result in withdrawal symptoms if the drug is discontinued. Hence compulsive drug use (addiction) is maintained to avoid the unpleasant withdrawal symptoms.

The central mechanism in this theory is homeostatic in nature. There are hundreds of examples of compensatory responses that operate in living systems. All mammals live most comfortably within a specific optimal range of values for variables such as core body temperature, blood composition, blood pressure, and others. Brain cells, for example, are very temperature-sensitive. Their electrophysiological responsiveness to stimulation changes dramatically with even small deviations from a core body temperature of 37 degrees Celsius.

Cells in the hypothalamus are sensitive to temperature deviations and send feedback signals into the peripheral nervous system to make compensatory adjustments to the rest of your body that are designed to bring body temperature back into the optimal range. If you're a member of the Polar Bear Club and just finishing a brisk winter swim in Lake Michigan, as you leave the water your hypothalamus will scream at your autonomic nervous system to make adjustments. It will send signals whose end results are to make you shiver (in an attempt to warm the muscles), develop goose bumps (to fluff your nonexistent fur), and turn your skin blue (a result of blood moving away from cold surface tissues to warm the inner sensitive core of the body).

If, on the other hand, you are involved in strenuous exercise on a very warm day, the hypothalamus activates systems to dissipate heat such as perspiration (which cools the skin by evaporation) and increasing blood flow to the skin surface, where heat can be radiated away (and make your face appear flushed). There are many other examples of homeostatic control by the hypothalamus and other brain regions, including the regulation of blood oxygen, volume, salinity, acidity, and so forth.

Ingestion of a drug that binds with brain receptors does the same thing as environmental changes do in the examples given above. It causes a shift in normal neurotransmission that will, in turn, elicit compensatory mechanisms that attempt to bring the system back to some rough homeostatic or allostatic level. But the compensatory response actually competes with the drug-induced response. This process results in users needing to increase their dosage to get the same effect. This is known as drug tolerance. When drug ingestion is stopped, the compensatory response is still active, so there is a net shift toward effects in the *opposite* direction of those induced by the drug.

These effects operate at a number of levels and comprise the symptoms associated with withdrawal. Thus at the sensory level, the pleasure induced by drug ingestion is combated by unpleasant opposing processes that are left unchecked during the withdrawal state.

Although the classic hedonic model is appealing for a number of reasons, experimental and observational findings suggest that it is limited in accounting for several aspects of the addictive process. One problem with the theory is that it fails to explain why individuals addicted to drugs often relapse into use even after they are free of withdrawal symptoms. The compensatory response that underlies withdrawal decays over time, and therefore the supposed prime reason for continued use no longer exists. Another major problem for the theory is that many addictive substances are not terribly pleasant at first, yet they still drive compulsive behaviors. If there is no hedonic value from the start, why would use continue beyond the first neutral or even negative experience? A typical example of this effect is first-time cigarette use, which most people find very unpleasant.

Aberrant learning theory is probably the newest perspective on addiction. The basic idea is grounded in associative learning theory (see chapter 3), where a stimulus and a response become paired. Experiments in rodents have shown that distinct parts of the brain become activated when an animal is rewarded with an addictive drug (for example, the nucleus accumbens, a collection of neurons in the forebrain) and often in anticipation (for example, the medial prefrontal cortex) of the reward. There are endogenous brain receptors for all psychoactive compounds. For instance, mu receptors are activated by morphine and heroin, and dopamine receptors are activated by cocaine and amphetamines. We, of course, do not have these agents circulating naturally through our bodies, but there are endogenous analogues that have evolved as neurotransmitters and neuropeptides. There are a number of ways one can measure the affinity or strength with which a natural neurotransmitter can activate a receptor and trigger the ensuing biochemical process. The aberrant learning theory posits that modern addictive drugs have much greater potency in activating endogenous receptors than their natural counterparts and hence act like an abnormally powerful stimulus. A neural system that is designed to learn stimulus-response pairings will be hyper-activated by such a powerful stimulus, and this learning might be very quick and enduring. The theory suggests, then, that the neural systems that typically regulate positive emotions such as pleasure are fooled by an abnormally powerful stimulus (a pure drug rather than the less potent endogenous counterpart) that essentially hijacks this normal process and creates a false fitness indicator signal.

One limitation of the aberrant learning theory is that there has never been a reasonable explanation as to why an unusually strong stimulus-response association would immutably lead to compulsive behaviors. Since it is still a relatively new perspective, more studies are needed to fill in some of the missing pieces that connect alterations at the neural level to ultimate changes in psychological functioning and behavior.

The third major theoretical perspective on addiction involves the loss of inhibitory control. This perspective is really a much broader theory about the origins of impulsive behaviors and extends well beyond addiction. The famous story of Phineas Gage, a man whose life changed in the blink of an eye, illustrates how damage to the brain areas that control impulsivity affects many behaviors that are also altered by addiction.

The year is 1848, and like much of New England, Vermont is slowly stirring from its sleepy agrarian pace toward industrializa-tion. Part of this transformation resides in the construction of several railways that will soon connect the major cities of the region. As one follows the planned route of the Rutland and Burlington line from Bellows Falls northward, the township of Duttonsville emerges within just a few miles, and we find ourselves abruptly shifting direc-tion, moving on a westward path toward Proctorsville. It is in this small town that we find Phineas P. Gage, a foreman contracted to lay ties for this segment of our new railway.

Gage is a great favorite with the men in his gang of navigators or navvies, a term left over from the early days when many of the laborers working on the railway found prior employment in the construction of canals used for transporting goods. The men revere him because he is a diligent worker possessing an iron frame and is scrupulously fair in the way he treats those in his employ. He does not play favorites, but rather allots tasks and pay in an equitable manner. This exacting and decisive nature also makes Phineas a favorite with his employers, who consider him "the most efficient and capable man" they have.

Indeed, everyone seems to like Phineas Gage. Friends and neighbors describe him as quiet and respectful of others, and such "temperate habits" are the hallmark of a good foreman. Bosses who fail to live up to these standards are unpopular, and within the culture of violence that persists among the rail workers at present, run the risk of being attacked and possibly killed. Indeed, by the time Gage is at work on the line, several foremen have already been fatally wounded by those in their charge in and around the Cavendish area.

At the moment, Phineas and his gang are laying a portion of the track that smacks snug up against the Black River. They will have to blast away large outcroppings of rock so the line can assume a level and straight flow. Blasting involves several stages that must be performed carefully and in the correct sequence. After a small diameter hole is drilled, a safety fuse is knotted and placed into the hole. Then an explosive powder, usually made from a mixture of sulfur, charcoal, and a nitrate such as saltpeter, is placed over the fuse. Finally, sand or clay is poured over the powder and compacted with a tamping iron. The purpose of tamping is to consolidate the explosive force to as small an area as possible, thereby allowing the charge to detonate into the rock with greater efficiency rather than escaping ineffectually back out the hole.

Phineas is an old hand at this and has had his tamping iron forged by a local blacksmith to his own specifications—"to please the fancy of the owner," others would later say. It is 3 feet, 7 inches long, 1¼ inches in diameter at the larger end, and tapered to a sharp point of

¼-inch diameter at the other end. It is quite hefty, in all weighing almost 13½ pounds.

The gang has been at work all day along a bend in the bank of the river, and Phineas has just poured explosive powder into a shallow hole about three feet deep. While waiting for his assistant to pour sand over the mixture before he tamps the charge, a commotion among the men erupts just a few feet behind him. Looking over his right shoulder, he turns to see that all is okay but must feel every ounce of the full weight of the tamping iron in his hand, for it has been a long day and it is 4:30 P.M., almost quitting time. As he returns his attention to tamp the charge, a shattering explosion occurs, and his iron, sharp side forward, is thrust up, piercing his left cheek. The projectile is moving with such extreme force that it penetrates the base of his skull, plowing through the front portion of his brain, and exits out the top of his head. Phineas immediately falls back to the ground, while his tamping iron, covered in blood, has rocketed an additional hundred yards before returning to earth. Amazingly he is still conscious and begins to speak to those stunned workers gathering around him.

Gage manages to stand awkwardly and "walks a few rods"—fifty feet or so—to an ox cart, where he rests against the foreboard and is driven the three-quarters of a mile back to his room in town. The cart lurches west past the intersection of Depot and Main streets, and when it arrives at the tavern of Mr. Joseph Adams, owner and solicitor, Phineas walks to the back, allowing two of his men to help him down. He then gently moves a short distance up three stairs and comes to rest in a small chair on the tavern's veranda to await medical attention. It is not until some two weeks later that Phineas emerges from his semiconscious state and begins to stir.

As improbable as the accident and recovery were, and as delighted in his progress as his physicians could be, Gage's friends soon began to realize that something in Phineas was amiss—"Gage is no longer Gage," several remarked. Within six weeks of the accident, much of the temperament of the young foreman working on the Rutland and Burlington Railroad has now changed. Those things that made Phineas who he was, at least to his friends and family, seem to have been stripped away with the shearing force of the tamping iron. Rather than being described as one who "possessed a well-balanced mind . . . looked upon by those who knew him as a shrewd, smart businessman, [and] . . . persistent in executing all his plans of operation," we find a new set of postaccident adjectives used to describe his character.

After recovering his physical strength, Phineas pleaded for his old job as foreman but was turned away. His contractors, who considered him the most efficient and capable man in their employ before the accident, regarded the change in his personality and behavior as so severe that they could not grant him his position again. In a letter to the Massachusetts Medical Society, Gage's physician, Dr. John Harlow, writes, "the equilibrium or balance, so to speak, between his intellectual faculties and animal propensities, seems to have been destroyed. He is fitful, irreverent, indulging at times in the grossest profanity (which was not previously his custom), manifesting but little deference for his fellows, impatient of restraint or advice when it conflicts with his desires, at times perniciously obstinate, yet capricious and vacillating, devising many plans of future operation, which are no sooner arranged than they are abandoned in turn for others appearing more feasible. A child in his intellectual capacity and manifestations, he has the animal passions of a strong man."

The description of the new Phineas as childlike was common among his friends, family, and even his physician. Gone was the man who routinely managed a large gang of navvies with competence and an efficiency that was the envy of other foremen. In his place we find a mercurial man who has no patience for plans or goals, little emotional attachment to previous friends, and seems to have lost the ability to anticipate the future and control his impulsive desire to live only in the present.

We now understand that the tamping iron damaged a large portion of Phineas's frontal cortex, an area that we know from experiments in animals and humans is intimately involved in higher cognitive functions, including the ability to balance present needs with longer-term consequences. Rats that have very restricted damage to key neural pathways that link brain areas involved in positive emotions such as the nucleus accumbens with the prefrontal cortex have impaired learning on a number of tasks. For instance, in one study paradigm rats are given a sweetened treat, but it is accompanied by a mild foot shock. Normal rats typically learn within one or two trials to avoid the treat (even though they would normally consume it readily) and the subsequent foot shock. Rats with damage to this inhibitory pathway, however, go back again and again for the treat despite the repeated foot shock that clearly causes them distress.

There is emerging evidence that chronic exposure to some addictive drugs such as amphetamines and cocaine can reduce neural activation in the frontocortical systems that seem to regulate inhibitory control. Consistent with these findings, addicts often exhibit a very similar pattern of deficits on neuropsychological tests to those observed in patients with damage to frontocortical systems. Taken together, this suggests that individuals who have weakened frontocortical systems may be less able to regulate impulsivity, and this limitation may contribute, in part, to drug-using behaviors.

The loss of inhibitory control and its associated behavioral problems seem to be robust phenomena in certain individuals, particularly those with obvious brain damage. It is still very unclear, however, if it is possible that more subtle deficits (perhaps not involving structural damage) in frontocortical functioning would predispose an individual to drug-seeking. It is very likely that a loss of inhibitory control is just one component of the addictive process and may work alongside other mechanisms to drive both drug-seeking and compulsive drug use.

The fourth major theory of addiction is the modern hedonic perspective. This view is rooted in fairly recent findings from a handful of neuroscientists who have shown that there seem to be different neural systems that regulate the "wanting" of a drug versus the "liking" of a drug. Kent Berridge and Terry Robinson, working at the University of Michigan, developed what is more formally called the "incentive sensitization" theory of addiction. In a series of elegant experiments, the researchers delineated two neural systems that contribute to the addictive process in fundamentally different ways.

Experiments in Berridge's lab and those of others have shown that in rats, addictive drugs alter the nucleus accumbens and related brain circuits that regulate motivational behaviors. If these circuits are lesioned in rats, the animals no longer display normal motivational behaviors such as seeking natural rewards (for example, sex, food, and water). Chemical activation of this circuitry when it is intact facilitates these motivational behaviors. This circuitry is part of the larger mesolimbic dopamine system that involves projections from brainstem regions to a number of pleasure centers such as the nucleus accumbens, larger striatum, and portions of the frontal cortex (see chapter 3).

For years, scientists thought this was the only brain system involved in reward, but we now know that at least four major systems are responsible for what we colloquially refer to as pleasure. Recent experiments have suggested that the mesolimbic dopamine system is primarily responsible for regulating motivational behaviors that make us "want" things. For instance, genetic manipulations that create mice with hyperactivated mesolimbic dopamine systems result in animals that are more motivated to obtain rewards and less distracted in reaching this goal than normal everyday mice. A major point, however, is that these animals do not display an enhanced "liking" of the reward once it is obtained—they don't consume any more than normal mice once they actually have the reward. The incentive sensitization theory holds that repeated use of addictive drugs enduringly alters this brain circuit, which in turn makes the drugs more desirable, resulting in a positive feedback loop. The process is analogous to activation of skin histamine receptors that causes us to scratch, thereby leading to further histamine release.

The mesolimbic dopamine system is only part of the hedonic story. The other major component is the brain's opioid system. Injection of chemicals that boost opioid neurotransmission in the basal forebrain markedly increases the actual consumption of palatable foods in rats. Moreover, opioid drugs given to humans and rodents increase their hedonic reactions to sucrose. Working in Berridge's laboratory, Susana Pecina has identified several "hedonic hot spots" in rats that when activated by opioid agonists enhance their natural pleasurable response to sucrose. Likewise, blocking opioid neurotransmission at these sites decreases the hedonic response or "liking" of sucrose.

These findings address a key limitation of the traditional hedonic model, namely, why people become addicted to an initially unpleasant experience (such as smoking or ingesting alcohol for some individuals). Indeed, many people who are addicted report that the pleasurable feelings initially felt when first using the drug subsided over time with continued use, yet they still felt strongly compelled to use the drug—they still wanted it, even though they did not necessarily like it. The fact that there are two entirely different neural systems that mediate distinct components of the pleasure instinct helps explain why many times there seems to be a dissociation between wanting a drug and liking a drug.

These components may have very different dynamics during the different states of drug use. Initial drug use, for instance, is most likely mediated by both mesolimbic dopamine circuits and the opioid system (that is, both wanting and liking). The transition to compulsive drug abuse may, in turn, be mediated primarily by the mesolimbic system, since many addicts complain that they do not experience the same hedonic response or "liking" once addicted as they initially did from the drug. One can further imagine that overactivation of the mesolimbic dopamine system, coupled with disruption to fronto-cortical inhibitory circuitry that may occur with amphetamine and cocaine use, could result in a particularly dangerous combination where drug wanting is enhanced and inhibitory control is reduced.

Berridge's lab has recently discovered that yet another brain system may help regulate "liking." The cannabinoid system (see chapter 6) overlaps in many locations with the opioid system. For instance, both have connections in the nucleus accumbens, but, of course, utilize different chemical neurotransmitters. Microinjection of the cannabinoid agonist anandamide into the nucleus accumbens of rats enhances their liking responses to sucrose in the same way as do opioids. Further work is now being done in multiple laboratories to determine how broad the cannabinoid circuitry is and to what extent it overlaps with the other two transmitter systems involved in the pleasure instinct.

Each of these three systems is known to have wide networks that stretch across the entire brain. In addition to the forebrain locations mentioned above, each of the neurotransmitter systems has brain-stem sites that seem to play a similar role in the wanting or the liking component of the pleasure instinct. Indeed, a fourth neurotransmitter system may involve benzodiazepine/GABA. Work in the late 1980s demonstrated that when decerebrate animals that have had their brain-stem transected from the rest of the brain were injected with a benzodiazepine drug (which enhances GABAergic neurotransmission), it enhanced their liking reactions to sweet tastes.

Hence there may be four distinct neurotransmitter systems that mediate the pleasure instinct, each with brain-stem origins. The fact that there could be so many brain-stem locations for inducing different components of the pleasure instinct is interesting and expected since, as we have seen in earlier chapters, these areas are the first brain regions to develop during gestation. Indeed, proper functioning of brain-stem areas is critical, since they control so many fundamental processes that support living organisms (for example, respiration, sleep-wake cycles, feeding, thermal regulation, and so forth). These are ancient brain regions, conserved across all mammalian species. Evolution has built upon their foundation. The central theory espoused in this book is that these brain-stem sites are responsible, in part, for using pleasure to nudge developing humans toward certain stimuli that must be experienced for normal brain development to continue. We have described this process as a kind of neural bootstrapping where activation of key brain regions induces pleasure when the right kinds of stimuli are experienced. The processing of these stimuli is required to stimulate the development of "higher" brain areas. Such a mechanism is dependent on the ability to activate the pleasure instinct at a number of brain regions as the developmental process unfolds.

The pleasure instinct is regulated by several major brain systems that originate in the brain-stem and project to higher levels of the brain. Each system may support different components of the initiation, sensorial processing, and perceptual feeling of pleasure during normal functioning. This pleasure circuitry coevolved with potentially addictive substances in the forms of plant and fruit compounds, but we have no way of knowing whether such substances led to addictive behaviors in early *Homo sapiens*. It is clear, however, that the refinement of certain compounds into more potent forms has led to their pursuit via the pleasure instinct on a pathological scale. The jump from fructose and lactose to refined table sugar is one example. The invention of distilled alcohol and synthetic compounds are two more. These agents mimic endogenous neurotransmitters that activate systems that normally use the pleasure instinct as a common metric for evaluating fitness. A key problem with modern compounds (especially those that are synthetic) is that they have tremendous potential for overactivating these circuits, since they are not metabolized in the same manner as their naturally occurring counterparts.

Although the modern hedonic theory of addiction has addressed some limitations of the earlier hedonic model, there are still many open questions about how addiction occurs and what can be done therapeutically to intervene early. The perspective advanced here does shed some light on several issues. First, it provides a biologically based mechanism, consistent with evolutionary theory, for understanding why we are all susceptible to drug and alcohol abuse. The same neural mechanisms that promote normal brain development through birth and well into the adolescent years drive us toward certain stimuli that signal high fitness value. This biological imperative translated into improved odds for surviving to reproductive age in ancestral environments. Clearly, however, this process can lead us toward some stimuli and experiences that may have adverse consequences in our modern environment.

The theory suggests, then, that each of us—not just the few with an unlucky genetic disposition—is susceptible to addiction. From this perspective, it is not so surprising that many people become addicted to these substances. Indeed, it is perhaps most surprising that only a relatively small percentage of people who experience drugs actually become addicted. Of course, a great deal of variation exists from individual to individual in how these neural systems communicate with one another, and some of this is undoubtedly modulated by genetic influences. Much of the evidence discussed above indicates that these systems are also readily modulated by environmental experiences, including drug use, stress, and a host of other life circumstances.

Even a casual observer would note the persistent relationship between chronic stress and drug use. Chronic stress and the associated activation of the body's response (for example, increased cortisol, adrenocorticotropin hormone, and corticotropin-releasing factor) have severe and deleterious effects on the neural systems that regulate emotions, including the pleasure instinct. Often these neural changes accumulate slowly, but they can last long after the stress has been reduced or even eliminated. For instance, in a landmark study, ethologist Dee Higley and colleagues showed that adult rhesus monkeys that were stressed for six months immediately after birth by being removed from their mothers exhibited increased stress responses (measured both physiologically and behaviorally) when compared to littermates who were allowed parental attachment during this critical period. Interestingly, the adult monkeys who were stressed at birth exhibited increased ethanol consumption under free-range conditions when compared to normally reared littermates. Monkeys that were not permitted to form a social attachment with their mothers during this critical developmental period grew up to be adults that exhibited greater fear and startle responses and increased physiological markers of stress (for example, cortisol production), and were more likely to consume a mind-altering substance than those that were able to establish a maternal bond during this period.

This is consistent with my experiences at the residential treatment house. Perhaps the most intriguing part of working at this facility was that the kids seemed to use drugs that elicited specific portions of the pleasure circuitry that compensated for the type of stress they were experiencing. Alberto, like many gang members, used methamphetamines to activate his mesolimbic and associated neural systems, giving him a sense of high energy, confidence, and strength, all requirements to combat the stress that accompanied gang life. Christine, on the other hand, gravitated toward increased sex and heroin use, which strongly activated her opioid system, eliciting a sense of calm and serenity—feelings that compensated for the abrupt loss of social attachment to her parents and grandparents. When adequate social bonds failed to develop (signaling a fitness decrement), Christine showed a tendency to engage the neural systems that promote feelings of attachment and calm through natural (via increased sex) and pharmacological means.

Understanding how the pleasure instinct may contribute to drug-seeking, drug use, and addiction is important, since the theory has implications for treatment. For instance, much of the literature examining the biology of addiction has concentrated on the meso-limbic dopamine system. Drug use can alter this system in a number of ways, including (1) decreases in dopamine-containing cell size in the basal forebrain; (2) enduring increases in postsynaptic dopamine receptor sensitivity; (3) alterations in the release of dopamine from presynaptic sites; and (4) a cascade of intracellular changes that occur, which ultimately impact dopamine transmission. Each of these presents a possible therapeutic target for combating drug-seeking and addiction on the "wanting" side. However, our theory predicts that agents that downregulate dopamine transmission may also have anti-hedonic effects, perhaps making them less tolerable to patients.

When we consider experiences like those of Alberto and Christine, and add what is now understood about the different neural systems involved in the pleasure instinct, it is probably safe to conclude that not all drugs induce the same hedonic experience. Another potential therapeutic target might be the opioid system, using compounds such as oxytocin agonists and prolactin agonists, which have been shown to reduce separation distress in animal models. Clonidine, an alpha-1 noradrenergic agonist, has been shown to reduce separation distress in rats and is already being used effectively in clinical practice to amerliorate opioid withdrawal.

Clearly, far better than treating an addiction already in progress is to implement ways to limit the likelihood of people feeling disenfranchised from family and society in the first place. There is accumulating data showing strong relationships among disruptions in normal social dynamics, drug use, and abuse. There is also a growing scientific literature demonstrating that both drug use and social bonding engage common pleasure circuitry. Providing additional social support structures, particularly for adolescents, might help reduce the likelihood of engaging these circuits through drugs and alcohol if they are experiencing social problems.

Finally, it should be said that the pleasure instinct offers but a single perspective on addiction. Certainly, no one theory explains all of addiction. Given what we now know about the disparate brain systems that different drugs engage, it is highly unlikely that there will ever be a single cure that fits all. However, as basic biological research continues to unveil the mysteries of how emotions are modulated by brain circuits, each new discovery presents a potentially novel therapeutic target. Understanding the environmental and experiential factors that also engage these circuits might offer therapeutic avenues that are just as appealing (or perhaps even more so) as those dreamed up by pharmaceutical companies. As we will see in the remaining chapter, modern life offers us a plethora of nonpharmaceutical ways to engage the neural circuits controlling the pleasure instinct.

Chapter 12

Parsing Pleasure

Though sages may pour out their wisdom's treasure,
there is no sterner moralist than Pleasure.

—Lord Byron, *Don Juan*

All the labor of man is for his mouth, and yet the
appetite is not filled.

—Ecclesiastes

This book began with a single question: why does pleasure exist? A passionate debate, framed on one side by hedonists and on the other by stoics, has raged since antiquity about the nature and utility of pleasure, and has influenced virtually all facets of social life in the Western world. Throughout the course of history popular opinion on the matter of pleasure—intermixed with its many political, legal, religious, and moral implications—has never found a stable resting point. Instead, it has swung back and forth like a pendulum. Conservative eras in which the pursuit of pleasure was, at least publicly, tempered have often been followed by more permissive periods, and vice versa.

Epicureans are usually thought of as representing an extreme position in this debate, one that advocates unchecked pleasure-seeking to the abandonment of all social responsibilities. Yet this is a modern distortion of Epicureanism, which originally held that happiness is attainable only if one distinguishes between pleasures that are natural and necessary, such as eating and drinking, from those that are simply desired. Ancient Epicureans followed a rather austere lifestyle, sacrificing some pleasures to avoid greater displeasures.

This philosophy laid the groundwork for the emergence of the sensualist ethics movement of the seventeenth and eighteenth centuries, a view that prevailed in the writings of thinkers such as John Locke and David Hume in England, and François Diderot in France. The central idea of this movement is that our senses are the final judge of what is positive or negative. Experiences that bring pleasure to the senses are deemed good, while those that displease are judged as bad. In this context the foundation for morality resides with the individual and is relativistic, a viewpoint that ultimately gave rise to utilitarianism in the nineteenth century.

This book entertains a third point of view that is distinctly different from those of the hedonists, who equate pleasure with happiness, and the modern-day stoics, who are guided by varying belief systems, yet at their core share an uneasy truce with pleasure. All of our emotions, including the experience of pleasure, have been shaped by natural selection to cope with challenges and opportunities that have recurred over the course of hominid evolution. In this context, pleasure may point us down a particular path, but it is not the final destination.

From a strict evolutionary perspective, the story of pleasure is often summed up in a few pages as one dealing almost exclusively with sexual behavior and reproduction. Sex has been linked to pleasure through natural selection to maximize genetic propagation—end of discussion. As we've seen, however, the role of pleasure in human evolution and development is far more complicated and interesting.

The modern hedonic palate is amazingly diverse and shaped by a host of personality and cultural factors. The feeling of pleasure can be elicited by a wide range of experiences, each involving disparate sensory systems. Likewise, the expression of pleasure in terms of overt behaviors or verbal descriptions varies considerably within different contexts and cultural norms. Although each of us has unique likes and dislikes, there is a common core of sensations that the vast majority of humans label as pleasurable. Our preference for these core sensations arose though natural selection to ensure that we experience key sensory events during our early years that are *required* for normal brain development. This does not mean that we are all destined to be attracted to exactly the same phenomena, but rather to the same general classes of stimulus features. To really understand the choices we make in contemporary life, it is critically important to consider the modern problem of how cultural influences interplay with nature's program of sensory and behavioral preferences.

Addiction, for example, is certainly not a modern problem. There are historical accounts of addiction to everything from chocolate to sex to war. As we have seen, at its root addiction is a biological phenomenon regulated by the brain circuitry that is involved in reward and motivated behaviors (see chapter 11). Clearly, however, cultural norms have a way of insinuating themselves into the biological mechanisms that control these behaviors.

Modern addiction—whether we are talking about heroin, alcohol, nicotine, over-the-counter drugs, gambling, or your favorite soap opera—is often said to be facilitated by two key elements: the ease with which a drug or experience is acquired, and the rate at which it activates the central nervous system. Smoking, for instance, is highly addictive because tobacco products are readily available, and nicotine rapidly activates the dopaminergic brain systems we discussed in chapter 11. A third element that should be considered is whether the focus of an addiction involves more than one hedonic preference. Many addictions involve compounded stimuli. One can think of these as complex experiences that are comprised of two or more fundamental hedonic preferences. Smoking and alcohol are both addictive in themselves, but are far more addictive when combined together. Indeed, becoming addicted to either smoking or drinking alcohol markedly increases the odds of becoming addicted to the other. This makes intuitive sense, since both forms of stimulation are activating the same brain reward mechanisms.

As we have seen, nature's list of hedonic preferences is long and taps into every sensory domain. This provides an enormous palate of sensory experiences that can be combined in culturally acceptable ways to form compounded stimuli. Toddlers are not terribly burdened by cultural norms, so the impact of compounded stimuli is most apparent in the things that grab their attention. The appeal of shows such as *Barney and Friends* and *The Teletubbies* is easily understood when considered from this perspective. Both shows deliver daily doses of compounded stimuli that are sharply tuned to match several of the hedonic preferences we have been discussing. Included among these are stimuli formed by sharp contrasts, bright primary colors, pronounced lateral symmetry, repetitive sounds and movements, and exaggerated intonation and facial images.

Adults are not exempt from the lure of compounded stimuli, but their attraction to these experiences is shaped significantly by cultural expectations and previous life events. Memory and associative learning link everyday stimuli to primary reinforcers, and this process plays a large role in determining what an individual finds pleasurable.

In the past three chapters we saw how the pleasure instinct crafts time-sensitive developmental preferences for certain types of stimuli that, in turn, can emerge as receiver biases in adults. There is probably no better place to look for examples of this process than the modern consumer world. In 1957, Vance Packard published his classic book *The Hidden Persuaders*. This was the first broadly read account of how the advertising community was taking advantage of discoveries in a number of scientific disciplines to tap into subconscious processes for the benefit of product positioning.

Packard's book focused on psychological processes that might be important for making a product appear more desirable to a targeted consumer. Receiver biases may also make certain products seem more appealing; however, the mechanism can be clearly linked to developmental preferences forged by natural selection and, in some cases, magnified by sexual selection.

Symmetry, proportion, and rhythm are particularly good examples of receiver biases that clearly have a home in the advertiser's toolbox. Our innate preference for these features has been studied by marketing scientists for decades and applied to everyday commercials. For instance, symmetry is a particularly ubiquitous feature of design logos for consumer products ranging from automobiles to home electronics, where the implicit message is one of—you guessed it—fitness. Indeed, virtually every major manufacturer of consumer goods today employs or consults sensory scientists to optimally align the aesthetic features of a product with Gestalt forms that are innately pleasing. These scientists spend a great deal of time getting customer feedback on potential product design, packaging materials, promotional displays, advertising placement, and so forth. These days I can no longer go into a department store and simply shop. Instead, I find myself counting the many ways I am subtly manipulated by the growing armamentarium of advertising/marketing devices supplied by modern sensory science, evolutionary biology, and neuroscience.

In chapter 3 we discussed the pioneering work of psychologists James Olds and Peter Milner, who in 1954 were the first to discover the so-called pleasure centers of the brain. Although modern brain theorists are still grappling with what is actually experienced when these sites are stimulated, an important finding is that a fairly restricted and well-defined set of circuits has been implicated again and again in literally thousands of studies. There are many ways to promote activation of these circuits via different sensory domains and experiences. Similarly, once these circuits have been activated, they may signal a diverse number of other brain regions that are responsible for the appropriate behavioral response.

Today it is unclear whether we should call the limbic circuitry involved in reward and motivation "pleasure centers." Olds and Milner stayed away from this terminology initially, but early theorists simply could not resist. Normal functioning of this limbic circuitry is implicated in several processes that seem to be related to pleasure. These include reinforcement learning through reward, consummatory behaviors such as eating food and mating, and appetitive or motivated behaviors such as those leading up to the point of actually receiving the reward.

The portions of the brain where this circuitry resides are subcortical structures that have existed at least as long as the birth of mammals. The evolution of primates is a story that in many ways began with the limbic system and the behaviors/functions it supports. Nature rarely reinvents itself. The general pattern of brain circuitry one sees in limbic structures such as the hippocampus (for example, recurrent excitatory synapses modulated by feed-forward inhibition) forms the basis for much of what is seen in neocortical structures. The gradual emergence of the primate line was accompanied by neocortical development that was based on mammalian limbic circuitry.

This evolutionary sequence is echoed in the embryonic growth of humans. Each milestone of brain development that is traversed by the growing child is a reprise of our evolutionary past. Limbic structures give rise to neocortical structures, and, most importantly, *limbic functions give rise to neocortical functions*. The basic emotions, including pleasure, laid the foundation for the gradual emergence of higher cognitive processes unique to human primates (for example, language and musicality). Parents see this process unfold daily as limbic functions such as finding pleasure in primary colors, faces, symmetry, repetitive movements, and certain sounds direct babies and toddlers toward the necessary experiences that will guide the continual development of more complex functions that require a healthy neocortex.

With only twenty-five thousand or so genes constituting the entire human genome, nature is forced to conserve at every step. Developmental bootstrapping is an elegant compromise that engenders the individual with a rapidly growing brain and a built-in desire to seek experiences that will fine-tune this circuitry. This process has quite a bit of freedom and flexibility. It's somewhat akin to a robotics engineer getting the basic wiring and simple functioning working and then letting the machine do the rest of the programming based on a few simple rules. The first rule is that certain classes of stimuli must be experienced during development to ensure that the programming is continued. The second rule is that the particular stimuli the robot encounters in these general classes will fine-tune the machine to life in a specific environment.

Machines are generally not designed this way, of course, but biological organisms possess a delicate balance of preprogrammed growth that depends on experiential refinement. This freedom and flexibility has its costs. Addiction is the most obvious example of how bootstrapping as a general developmental process can lead to potential problems. Compounded stimuli are not just found in modern environments, but the ease with which so many different hedonic preferences can be satisfied at a moment's notice is clearly an oddity of contemporary life. In the twenty-first century we have extraordinary technological means to quench our thirst for pleasure, and the potency of addiction will increase when our drug of choice is mixed with other hedonic preferences.

These ideas and findings suggest a novel treatment strategy for dealing with addiction in the broadest sense. Focusing solely on the primary point of addiction (for example, smoking or consuming alcohol) will generally be less effective than identifying all of the hedonic preferences associated with the behavior and treating the compounded stimuli as a whole. This perspective suggests, for example, that treatment for alcohol addiction should also include a therapeutic means to reduce smoking (if applicable) and curtail any increases in sugar consumption. Such a scenario is typically not part of alcohol treatment programs, many of which allow (or even encourage) substituting one addictive substance for another—sweets in place of nicotine, caffeine in place of tobacco, and so forth. People often find it easier to treat addiction through substitution. You probably know at least one person who quit smoking or drinking alcohol, only to increase another behavior that is equally addictive. This happens with great frequency, since all hedonic preferences—regardless of the sensory domain (for example, taste, touch, vision, etc.)—stimulate the same brain circuitry involved in reward and motivation. Activity in this circuitry is regulated by a number of additional systems, with the aim of keeping it within a range of acceptable levels (allostasis).

Addiction is really just the tip of the iceberg. Understanding the role of pleasure in the modern context requires stretching back in historical time and considering its function in the evolution of our species. It also requires thinking about the role of pleasure in human brain development—from embryo through the toddler stages, into adolescence, and then adulthood. Understanding how pleasure influences our behavior is critical for parents who want to make intelligent choices about the growth environments they provide for their children. Yes, it is obvious that children need stimulation to develop along a normal trajectory, but not all stimuli are created equal. There are clearly optimal forms of stimulation that parents should bring into their children's world. The pleasure instinct will do the rest by causing the child to repeatedly engage in play with objects that have these features.

There is, of course, a tremendous benefit for adults as well. Pleasure has varied roles in our modern lives, many of which are influenced by its original developmental purpose. Understanding the original role of pleasure helps to frame the modern experience by revealing key features of our lives that it impacts, ranging from addiction to the influence of advertising on consumerism. Hedonic preferences do not simply go away as we grow older. While it is true that they become increasingly shaped by our experiences and cultural norms with age, they are still with us, anchored beneath all the top layers. There is no doubt that brain development hits its zenith in our early years, but it is a continual process throughout the entire life span. Old neurons die, new neurons are born. Since brain development slows significantly after adolescence, the role of pleasure in guiding this growth becomes less central. Nevertheless, the pleasure derived from experiencing the many forms of developmentally important sensations persists—a common gift shared by us all, yet flavored by our unique life events and conditions.

Notes

1 Foibles and Follies

6 *Until recently, scientists have*   For an excellent review, see chapters in Ekman, P, Davidson, RJ (1997) *The Nature of Emotion: Fundamental Questions*. Oxford University Press, Oxford, England.

6 *Academics have typically shied*   But see Johnston, VS (1999) *Why We Feel: The Science of Human Emotions*. Perseus Publishing, New York, NY.

7 *Sociobiology and its heir apparent*   Excellent introductions to evolutionary psychology include Konner, MJ (2001) *The Tangled Wing: Biological Constraints on the Human Spirit*. W.H. Freeman and Company, New York, NY; Barkow, JH, Cosmides, L, Tooby, J (1997) *The Adapted Mind: Evolutionary Psychology and the Generation of Culture*. Oxford University Press, Oxford, England; Buss, DM (1998) *Evolutionary Psychology: The New Science of Mind*. Pearson Allyn and Bacon, New York, NY; Pinker, S (1999) *How the Mind Works*. W.W. Norton and Company, New York, NY.

7 *Pleasure, as we shall see*   Cabanac, M (1992) Pleasure: the common currency. *Journal of Theoretical Biology* 155, no.2: 173–200.

7 *Such biases persist*   A positive reinforcer is a stimulus that serves to increase the likelihood of the behavior that produces it.

2 How to Win Friends and Influence People

12 *In his book* The Language Instinct   Pinker, S (1994) *The Language Instinct*. Perennial, New York, NY.

14 *Such increasingly complicated*   Humphrey, N (1992) *A History of the Mind*. Simon and Schuster, New York, NY; Mithen, S (1996) *The Prehistory of the Mind: The Cognitive Origins of Art, Religion, and Science*. Thames and Hudson, England; Klein, RG (2002) *The Dawn of Human Culture*. John Wiley and Sons, New York, NY.

16 *Many scholars agree*   Diamond, JM (1992) *The Third Chimpanzee: The Evolution and Future of the Human Animal*. Harper Collins, New York, NY; Deacon, T (1997) *The Symbolic Species*. W.W. Norton, New York, NY.

17 *In cognitive science circles*   For a recent review of the representation problem, see Pylyshyn, ZW (2002) Mental imagery: in search of a theory. *Behavioral Brain Sciences* 25, no.2: 157–182.

17 *But this is premature*   Example used by Hauser, MD (2000) *Wild Minds: What Animals Really Think.* Henry Holt and Company, New York.

18 *How did proto-emotions*   See Dunbar, R (1998) *Grooming, Gossip, and the Evolution of Language.* Harvard University Press, Cambridge, MA; Deacon, T (1997) *The Symbolic Species.* W.W. Norton, New York, NY; Pinker, *The Language Instinct.*

18 *We will employ a more modern*   See Striedter, GF (1998) Progress in the study of brain evolution: from speculative theories to testable hypotheses. *The Anatomical Record (New Anatomy),* 253: 105–112; and relevant chapters in Gould, SJ (2002) *The Structure of Evolutionary Theory.* Harvard University Press, Cambridge, MA.

18 *"The human race began to talk*   Cited in Locke, *The Child's Path to Spoken Language.*

20 *In humans this exchange*   For a review, see ibid.; Owens, RE (2000) *Language Development: An Introduction.* Pearson Allyn and Bacon, New York, NY.

20 *Detailed experiments following*   Stern, DN, Spieker, S, Barnett, RK, MacKain, K (1983) The prosody of maternal speech: infant age and context related changes. *Journal of Child and Language,* 10(1): 1–15; Halle, PA, de Boysson-Bardies, B, Vihman, MM (1991) Beginnings of prosodic organization: intonation and duration patterns of disyllables produced by Japanese and French infants. *Language and Speech,* 34, 299–318.

20 *"Communication is successful"*   See Locke, *The Child's Path to Spoken Language.*

22 *For instance, synthesized sounds*   Murray, IR, Arnott, JL (1993) Toward the simulation of emotion in synthetic speech: a review of the literature on human vocal emotion. *Journal of the Acoustical Society of America,* 93(2): 1097–108; Collier, WG, Hubbard, TL (2001) Musical scales and evaluations of happiness and awkwardness: effects of pitch, direction, and scale mode. *American Journal of Psychology,* 114(3): 355–375.

22 *Developmental psychologist Andrew Meltzoff*   For an overview of these experiments see Gopnick, A, Meltzoff, AN, Kuhl, PK (1999) *The Scientist in the Crib: Minds, Brains, and How Children Learn.* William Morrow, New York.

23 *Studies have also shown that infants*   For a recent review of this literature see Turati, C, Simion, F, Milani, I, Umilta, C (2002) Newborns' preference for faces: what is crucial? *Developmental Psychology,* 38(6): 875–882.

3 What Makes Sammy Dance?

26 *By the time he was twenty-four years old*   Published reports of case studies typically use a patient's initials or some other descriptor in an effort to maintain their anonymity. Heath, RG (1972) Pleasure and brain activity in man. Deep and surface electroencephalograms during orgasm. *Journal of Nervous Mental Disorders,* 154(1): 3–18.

26 *Impressed by the work of Olds*   Olds, J, Milner, P (1954) Positive reinforcement produced by electrical stimulation of septal area and other regions of rat brain. *Journal of Comparative Physiology and Psychology,* 47(6): 419–427.

26 *"They function in an almost continuous*   Quoted in Hooper, J, Teresi, D (1986) *The Three-Pound Universe.* Macmillan Publishing Company, New York.

26 *Electrodes and cannulas*   Olds, J, Milner, P (1954) Positive reinforcement produced by electrical stimulation of septal area and other regions of rat brain. *Journal of Comparative Physiology and Psychology,* 47(6): 419–427.

29 *In the fifty years since*   For recent reviews see Berridge, KC (2002) Pleasure of the brain. *Brain and Cognition,* 52(1): 106–128; Wise, RA (2002) Brain reward circuitry: insights from unsensed incentives. *Neuron,* 36(2): 229–240; Kelley, AE, Berridge, KC (2002) The neuroscience of natural rewards: relevance to addictive drugs. *Journal of Neuroscience,* 22(9): 3312–3320.

29 *As the circuit continued*   Olds, J (1976) Brain stimulation and the motivation of behavior. *Progress in Brain Research,* 45: 401–426; Berridge, KC (2002) Pleasure of the brain. *Brain and Cognition,* 52(1), 106–128; Wise, RA (2002) Brain reward circuitry: insights from unsensed incentives. *Neuron,* 36(2): 229–240; Kelley, AE, Berridge, KC (2002) The neuroscience of natural rewards: relevance to addictive drugs. *Journal of Neuroscience,* 22(9): 3312–3320.

30 *Behavioral scientists have used*   Woodward, DJ, Chang, JY, Janak, P, Azarov, A, Anstrom, K (2000) Activity patterns in mesolimbic regions in rats during operant tasks for reward. *Progress in Brain Research,* 126: 303–322.

31 *Understanding how associative learning*   For examples see Eichenbaum, H (2001) The hippocampus and declarative memory: cognitive mechanisms and neural codes. *Behavioral Brain Research,* 127(1–2): 199–207; Hasselmo, ME, McClelland, JL (1999) Neural models of memory. *Current Opinion in Neurobiology,* 9(2): 184–188; Wallenstein, GV, Eichenbaum, H, Hasselmo, ME (1998) The hippocampus as an associator of discontiguous events. *Trends in Neuroscience,* 21(8): 317–323; Kesner, RP, Gilbert, PE, Wallenstein, GV (2000) Testing neural network models of memory with behavioral experiments. *Current Opinion in Neurobiology,* 10(2): 260–265.

31 *Rats often learn to avoid*   Parker, LA (2003) Taste avoidance and taste aversion: evidence for two different processes. *Learning and Behavior,* 31(2): 165–172; Maren, S (2001) Neurobiology of Pavlovian fear conditioning. *Annual Review of Neuroscience,* 24: 897–931.

32 *"The dancing chicken is exhibiting"*   Breland, K, Breland, M (1961) The misbehavior of organisms. *American Psychologist,* 16: 681–684.

32 *In fact, sugars rated*   Sclafani, A, Fanizza, LJ, Azzara, AV (1999) Conditioned flavor avoidance, preference, and indifference produced by intragastric infusions of galactose, glucose, and fructose in rats. *Physiology and Behavior,* 67(2): 227–234.

37 *Others are activated by stress, learning*   For a review see Wallenstein, GV (2003) *Mind, Stress, and Emotions: The New Science of Mood.* Commonwealth Press, Boston.

43 *At present, there are several controversial*   Slezak, M, Pfrieger, FW (2003) New roles for astrocytes: regulation of CNS synaptogenesis. *Trends in Neuroscience,* 26(10): 531–535; Adams, P, Cox, K (2002) A new interpretation of thalamocortical

circuitry. *Philosophical Transactions of the Royal Society of London B Biological Sciences,* 357(1428): 1767–1779; Grossman, AW, Churchill, JD, Bates, KE, Kleim, JA, Greenough, WT (2002) A brain adaptation view of plasticity: is synaptic plasticity an overly limited concept? *Progress in Brain Research,* 138: 91–108.

46 *From culture to culture* Sugarman, L (2001) *Lifespan Development: Theories, Concepts and Interventions.* Psychology Press, New York.

4 The Pleasure of Touch

49 *By the time the world met him* Ruckel, I (2002) *Abandoned for Life: The Incredible Story of One Romanian Orphan.* JB Information Station Publishers, St. Louis.

50 *As late as 1915* Chapin, HD (1915) A plea for accurate statistics in children's institutions. *Transactions of the American Pediatric Society,* 27: 180.

50 *It is important to note* Holt, LE (1935) *The Care and Feeding of Children* (15th edition). Appleton-Century, New York.

51 *The pediatric wards of the famed Bellevue* Brennemann, J (1932) The infant ward. *American Journal of Diseases of Children,* 43: 577.

51 *"If they continued hurting themselves* Ruckel, I (2002) *Abandoned for Life: The Incredible Story of One Romanian Orphan.* JB Information Station Publishers, St. Louis.

53 *For touch to be perceived* Falk, D, Gibson, KR (2001) *Evolutionary Anatomy of the Primate Cerebral Cortex.* Cambridge University Press, Cambridge, England.

53 *Each of these sensory modalities* For a general review of the touch system, see Rowe, MJ, Iwamura, Y (2000) *Somatosensory Processing: From Single Neuron to Brain Imaging.* Taylor & Francis, New York.

54 *A very active area of research* Fox, K, Glazewski, S, Schulze, S (2000) Plasticity and stability of somatosensory maps in thalamus and cortex. *Current Opinion in Neurobiology,* 10(4): 494–497; Jones, EG (2000) Cortical and subcortical contributions to activity-dependent plasticity in primate somatosensory cortex. *Annual Review of Neuroscience,* 23: 1–37; Rauschecker, JP (2002) Cortical map plasticity in animals and humans. *Progress in Brain Research,* 138: 73–88.

54 *Psychologist William Greenough* Klintsova, AY, Greenough, WT (1999) Synaptic plasticity in cortical systems. *Current Opinion in Neurobiology,* 9(2): 203–208; Greenough, WT (1987) Mechanisms of behaviorally-elicited and electrically-elicited long-term potentiation. *International Journal of Neurology,* 21: 137–144.

54 *Experience-expectant stimuli* In Barnard, KE, Brazelton, TB (1990) *Touch: The Foundation of Experience.* International University Press, Madison, CT.

55 *Thus, whisker sensation* Kossut, M (1998) Experience-dependent changes in function and anatomy of adult barrel cortex. *Experimental Brain Research,* 123 (1–2): 110–116; Glazewski, S (1998) Experience-dependent changes in vibrissae evoked responses in the rodent barrel cortex. *Acta Neurobiologica Exp (Wars),* 58(4): 309–320; Fox, K (2002) Anatomical pathways and molecular mechanisms for plasticity in the barrel cortex. *Neuroscience,* 111(4): 799–814.

55 *For instance, when mice are placed*   For a review of these studies see van Praag, H, Kempermann, G, Gage, FH (2000) Neural consequences of environmental enrichment. *Nature Reviews Neuroscience,* 1(3): 191–198; Rosenzweig, MR, Bennett, EL (1996) Psychobiology of plasticity: effects of training and experience on brain and behavior. *Behavioral Brain Research,* 78(1): 57–65.

56 *Rather, the preferences for certain*   Victora, MD, Victora, CG, Barros, FC (1990) Cross-cultural differences in developmental rates: a comparison between British and Brazilian children. *Child Care Health and Development,* 6(3): 151–164; Chen, ST (1989) Comparison between the development of Malaysian and Denver children. *Journal of the Singapore Pediatric Society,* 31(3–4): 178–185; Lejarraga, H, Pascucci, MC, Krupitzky, S, Kelmansky, D, Bianco, A, Martinez, E, Tibaldi, F, Cameron, N, (2002) Psychomotor development in Argentinean children aged 0–5 years. *Pediatric and Perinatal Epidemiology,* 16(1): 47–60.

57 *Some recent studies have shown*   Eriksson, PS (2003) Neurogenesis and its implications for regeneration in the adult brain. *Journal of Rehabilitative Medicine,* (41 Suppl): 17–19; Kozorovitskiy, Y, Gould, E (2003) Adult neurogenesis: a mechanism for brain repair? *Journal of Clinical and Experimental Neuropsychology,* 25(5): 721–732; Eisch, AJ (2002) Adult neurogenesis: implications for psychiatry. *Progress in Brain Research,* 138: 315–342.

60 *For instance, children who have delayed*   Rine, RM, Cornwall, G, Gan, K, LoCascio, C, O'Hare, T, Robinson, E, Rice, M (2000) Evidence of progressive delay of motor development in children with sensorineural hearing loss and concurrent vestibular dysfunction. *Perceptual and Motor Skills,* 90(3 Pt 2): 1101–1112.

60 *The stress hormones*   See Wallenstein, GV (2003) *Mind, Stress, and Emotions: The New Science of Mood.* Commonwealth Press, Boston.

61 *Moreover, animals that are given*   For review see Wallenstein, GV (2003) *Mind, Stress, and Emotions: The New Science of Mood.* Commonwealth Press, Boston; Gilmer, WS, McKinney, WT (2003) Early experience and depressive disorders: human and non-human primate studies. *Journal of Affective Disorders,* 75(2): 97–113.

61 *For instance, gentle daily massage*   For a review see Fleming, AS, O'Day, DH, Kraemer, GW (1999) Neurobiology of mother–infant interactions: experience and central nervous system plasticity across development and generations. *Neuroscience and Biobehavioral Reviews,* 23(5): 673–685; Carvell, GE, Simons, DJ (1996) Abnormal tactile experience early in life disrupts active touch. *Journal of Neuroscience,* 16(8): 2750–2757.

61 *Although there have not been many controlled*   See articles in Barnard, KE, Brazelton, TB (1990) *Touch: The Foundation of Experience.* International University Press, Madison, CT.

62 *The twin who received the motion*   For a comprehensive review of early studies see Ottenbacher, KJ, Petersen, P (1984) The efficacy of vestibular stimulation as a form of specific sensory enrichment. Quantitative review of the literature. *Clinical Pediatrics (Phila.),* 23(8): 428–433.

5 In Praise of Odors

65 *This is true for humans, primates* Tanabe, T, Iino, M, Takagi, SF (1975) Discrimination of odors in olfactory bulb, pyriform-amygdaloid areas, and orbitofrontal cortex of the monkey. *Journal of Neurophysiology,* 38: 1284–1296; Tanabe, T, Yarita, H, Iino, M, Ooshima, Y, Takagi, SF (1975) An olfactory projection area in orbitofrontal cortex of the monkey. *Journal of Neurophysiology,* 38: 1269–1283; Pritchard, TC, Hamilton, RB, Morse, JR, Norgren, R (1986) Projections of thalamic gustatory and lingual areas in the monkey, Macaca fascicularis. *Journal of Comparative Neurology,* 244: 213–228; Zatorre, RJ, Jones-Gotman, M, Evans, AC, Meyers, E (1992) Functional localization and lateralization of human olfactory cortex. *Nature,* 60: 339–340.

65 *Experiments have shown that when humans* Zatorre, RJ, Jones-Gotman, M, Evans, AC, Meyers, E (1992) Functional localization and lateralization of human olfactory cortex. *Nature,* 60: 339–340; Small, DM, Jones-Gotman, M, Zatorre, RJ, Petrides, M, Evans, AC (1997) A role for the right anterior temporal lobe in taste quality recognition. *Journal of Neuroscience,* 17: 5136–5142; Small, DM, Jones-Gotman, M, Zatorre, RJ, Petrides, M, Evans, AC (1997) Flavor processing: more than the sum of its parts. *Neuroreport,* 8: 3913–3917.

66 *In her book* A Natural History Ackerman, D (1991) *A Natural History of the Senses.* Vintage Books, New York.

66 *One theory suggests that during the Devonian* Watson, L (2000) *Jacobson's Organ and the Remarkable Nature of Smell.* W. W. Norton & Company, New York.

69 *By the twenty-eighth week, Kai's placenta* Psychologist Julie Mennella first tested this idea by taking amniotic fluid samples from women who were undergoing a routine amniocentesis and had ingested either garlic or placebo capsules approximately forty-five minutes before the procedure. The odor of the amniotic fluid from women who consumed garlic was judged, by independent evaluators, as smelling stronger and more like garlic than the fluid from those who consumed only saline (Mennella, JA, Johnson, A, Beuchamp, GK [1995] Garlic ingestion by pregnant women alters the odor of amniotic fluid. *Chemical Senses,* 20: 207).

69 *In fact, scientists have speculated that odor* Eliot, L (2000) *What's Going on in There? How the Brain and Mind Develop in the First Five Years of Life.* Bantam Books, New York.

70 *Injections of simple saline solution* Smotherman, WP, Robinson, SR (1988) Behavior of rat fetuses following chemical or tactile stimulation. *Behavioral Neuroscience,* 102: 24–34.

70 *After birth, the rats that were exposed* Pedersen, PE, Blass, EM (1982) Prenatal and postnatal determinants of the first sucking episode in albino rats. *Development Psychobiology,* 15: 349–355.

70 *If their amniotic fluid is scented* Smotherman, WP (1982) Odor aversion learning by the rat fetus. *Physiology and Behavior,* 29: 769–771.

70 *Experiments performed in culturally diverse* Varendi, H, et al. (1996) Attractiveness of amniotic fluid odor: evidence of prenatal olfactory learning? *Acta Paediatrica,* 85: 1223–1227.

71 *Newborns also cry less and show* Varendi, H, et al. (1998) Soothing effect of amniotic fluid smell in newborn infants. *Early Human Development,* 51: 47–55.

71 *Within hours after birth, a breast-fed* Schaal, B (1988) Olfaction in infants and children: development and functional perspectives. *Chemical Senses,* 13: 145.

71 *Infants born to anise-consuming* Schaal, B, Marlier, L, Soussigan, R (2000) Human fetuses learn odours from their pregnant mother's diet. *Chemical Senses,* 25: 729–737.

72 *Researchers have found that newborns* Mennella, JA, Beauchamp, GK (1991) Olfactory preferences in children and adults. In *The Human Sense of Smell,* Lang, D, Dory, RL, Breipohl, W (eds.). Springer-Verlag, NY.

72 *Some of the most popular smells include strawberry* Schmidt, HJ, Beauchamp, GK (1988) Adult-like preferences and aversions in three-year-old children. *Child Development,* 59: 1138.

74 *Within four months* Russell, MJ, et al. (1980) Olfactory influences on the human menstrual cycle. *Pharmacology, Biochemistry, and Behavior,* 13: 737.

74 *The answer came* Empson, J (1977) Periodicity in body temperature in man. *Experientia,* 33: 342.

77 *Each person tested was brought* Wedekind, C, Furi, S (1997) Body odour preferences in men and women: do they aim for specific MHC combinations or simply heterozygosity? *Proceedings of the Royal Society of London B Biological Sciences,* 22: 1471–1479.

77 *In 2001, Wedekind's group* Milinski, M, Wedekind, C (2001) Evidence for MHC-correlated perfume preferences in humans. *Behavioral Ecology,* 12: 140–149.

6 For the Love of Chocolate

80 *Spouted, teapot-shaped vessels* Hurst, WJ, Tarka, SM Jr., Powis, TG, Valdez, F Jr., Hester, TR (2002) Cacao usage by the earliest Maya civilization. *Nature,* 418: 289–290.

82 *Placing a small amount of liquid* Barr, RG et al. (1994) Effects of intra-oral sucrose on crying, mouthing, and hand-mouth contact in newborn and six-week-old infants. *Developmental Medicine and Child Neurology,* 36: 608.

82 *The key to unraveling this mystery* di Tomaso, E, Beltramo, M, Piomelli, D (1996) Brain cannibinoids in chocolate. *Nature,* 382: 677–678.

84 *In a very real sense, the opioid* Molecular evidence has shown that the opioid system is highly conserved across anatomically diverse species and hence very old within the framework of hominid evolution. For instance, the gene sequence that codes the development of mu opioid receptors is essentially the same in humans, bovine chickens, bullfrogs, striped bass, thresher sharks, and the Pacific hagfish (Li, X, Keith, DE, Evans, CJ [1996] Mu opioid receptorlike sequences are present throughout vertebrate evolution. *Journal of Molecular Evolution,* 43: 179–184.)

84 *Study after study has shown that in humans* For examples, see Facchinetti, F et al. (1986) Hyperendorphinemia in obesity and relationships to affective state. *Physiology and Behavior,* 36: 937–940; Melchior, JC, et al. (1991)

Immunoreactive beta-endorphin increases after an aspartame chocolate drink in healthy subjects. *Physiology and Behavior,* 50: 941–944.

85 *At least two independent studies have found that obese* Facchinetti, F, et al. (1986) Hyperendorphinemia in obesity and relationships to affective state. *Physiology and Behavior,* 36: 937–940; Scavo, D, et al. (1990) Hyperendorphinemia in obesity is not related to affective state. *Physiology and Behavior,* 48: 681–683.

85 *Hence, attachment behaviors depend on opioid* Moles, A, Kieffer, BL, D'Amato, FR (2004) Deficit in attachment behavior in mice lacking the mu-opioid receptor gene. *Science,* 304: 1983–1986.

87 *These include tongue protrusions to reject* Steiner, JE (1973) The gustofacial response: Observation on normal and anencephalic newborn infants. In *Fourth Symposium on Oral Sensation and Perception.* JF Bosma (ed.), U.S. Department of Health, Education, and Welfare, Bethesda, MD.

88 *Studies performed during this time show that fetuses* Mistretta, CM, Bradley, RM (1975) Taste and swallowing in utero: a discussion of fetal sensory function. *British Medical Bulletin,* 31: 80–84.

88 *Likewise, rats born to mothers who consume* The implications of this learning process for understanding and treating addiction will be discussed in chapter 9.

89 *But this has changed with the discovery that simply putting a fatty* Mattes, RD (2001) The taste of fat elevates postprandial triacylglycerol. *Physiology and Behavior,* 74: 343–348.

90 *These findings are probably no surprise* Drewnowski, A, Greenwood, MR (1983) Cream and sugar: human preferences for high-fat foods. *Physiology and Behavior,* 30: 629–633.

90 *Michael Crawford of the Institute* Cunnane, SC, Crawford, MA (2003) Survival of the fattest: fat babies were the key to evolution of the large human brain. *Comparative Biochemistry and Physiology,* 136(1): 17–26.

92 *They suggest the most dramatic increase* Ibid.

92 *For instance, both ALA and LA* For a review, see Bourre, JM (2004) Roles of unsaturated fatty acids (especially omega-3 fatty acids) in the brain at various ages and during aging. *Journal of Nutrition, Health, and Aging,* 8(3): 163–174.

92 *Another study showed that ALA* Ibid.

93 *It is known that humans and other animals* See, for example, Imaizumi, M, Takeda, M, Sawano, S, Fushiki, T (2001) Opioidergic contribution to conditioned place preference induced by corn oil in mice. *Behavioral Brain Research,* 121(1–2): 129–136.

93 *For instance, if LA and sucrose* Gilbertson, TA (1999) The taste of fat. In *Pennington Center Nutrition Series: Nutrition, Genetics, and Obesity.* Louisiana State University Press, Baton Rouge, 192–207.

93 *In a series of behavioral experiments* Herzog, P, McCormack, DN, Webster, KL, Pittman, DW (2003) Linoleic acid alters licking responses to sweet, sour, and salt tastants in rats. Abstract, 25th Annual Meeting of the Association for Chemoreception Sciences.

93 *Likewise, when Pittman's rats* Ibid.

7 The Evolution of the Lullaby

97 *Indeed, given the omnipresence of music* Hauser, MD, McDermott, J (2003) The evolution of the music faculty: a comparative perspective. *Nature Neuroscience,* 6(7): 663–668; Trehub, SE (2003) The developmental origins of music. *Nature Neuroscience,* 6(7): 669–673; Miller, GF (2000) Evolution of human music through sexual selection. In Wallin, NL, Merker, B, Brown, S (eds.) (2001), *The Origins of Music.* MIT Press, Cambridge, MA, 329–360.

98 *The first group views music as an interesting* Pinker, S (1994) *The Language Instinct: How the Mind Creates Language.* William Morrow & Company, New York; Pinker, S (1997) *How the Mind Works.* W. W. Norton & Company, New York.

98 *This view, however, has its problems* Williams, GC (1966) *Adaptation and Natural Selection.* Princeton University Press, Princeton, New Jersey.

98 *Finally, a third school of thought* Miller, GF (2000) *The Mating Mind.* Doubleday, New York; Levitan, DJ (2006) *This Is Your Brain on Music.* Dutton, New York; Mithen, S (2006) *The Singing Neanderthals: The Origins of Music, Language, Mind, and Body.* Harvard University Press, Cambridge, MA.

99 *Indeed, adaptations driven by sexual* Zahavi, A (1997) *The Handicap Principle.* Oxford University Press, Oxford, England.

100 *This has led some theorists to focus on the similarities* Patel, AD (1998) Syntactic processing in language and music: Different cognitive operations, similar neural resources? *Music Perception,* 16: 27–42; Trehub, SE (2003) The developmental origins of musicality. *Nature Neuroscience,* 6: 669–673.

100 *They viewed music as being built from a hierarchy* Lerdahl, F, Jackendoff, R (1983) *A Generative Theory of Tonal Music.* MIT Press, Cambridge, MA.

102 *They also discriminate two melodies* Schellenberg, EG, Trehub, SE (1996) Natural musical intervals: evidence from infant listeners. *Psychological Science,* 7: 272–277.

103 *To date, this experiment has only been performed with rhesus* Trainor, LJ, Trehub, SE (1993) What mediates infants' and adults' superior processing of the major over the augmented triad? *Music Perception,* 11: 185–196.

104 *Practically everyone agrees on what is* Trehub, SE, Unyk, AM, Trainor, LJ (1993) Adults identify infant-directed music across cultures. *Infant Behavioral Development,* 16: 193–211.

105 *Interestingly, the abnormalities persisted* Chang, EF, Merzenich, MM (2003) Environmental noise retards auditory cortical development. *Science,* 300: 498–502.

105 *And when it finally gets them* Howard Hughes Medical Institute Newsletter (2003)—available online at http://www.hhmi.org//news/chang.html.

107 *Research has shown that fetuses older* Shahidullah, BS, Hepper, PG (1994) Frequency discrimination by the fetus. *Early Human Development,* 36: 13–26.

108 *Indeed, it is now clear that the sounds a fetus* Hepper, PG (1988) Fetal "soap" addiction. *Lancet,* June: 1347–1348.

8 In Search of Pretty Things

114 *In early primates this high concentration* Neurobiologist John Allman was the first to point out this shift from olfactory to visual dominance that occurred with the evolution of primates. For a description, see Allman, J (1977) Evolution of the visual system in early primates. In *Progress in Psychobiology and Physiological Psychology,* ed. Sprague, J and Epstein A. Academic Press, New York, 1–53.

115 *Evolutionary biologists have argued* Cartmill, M (1972) Arboreal adaptations and the origin of the order primates. In *The Functional and Evolutionary Biology of Primates,* ed. Tuttle, R. Aldine-Atherton Press, Chicago, 97–212; Martin, R (1990) *Primate Origins and Evolution: A Phylogenetic Reconstruction.* Princeton University Press, Princeton, NJ.

115 *This process resulted in a higher density* Nathans, J (1989) Genes for color vision. *Scientific American,* 260: 42–49.

115 *Recent work has shown that the advance from dichromacy* Osorio, D, Vorobyev, M (1996) Colour vision as an adaptation to frugivory in primates. *Proceedings of the Royal Society of London,* B, 263: 593–599.

118 *The latter stream is known as the "what"* Ungerleider, LG, Haxby, JV (1994) "What" and "where" in the human brain. *Current Opinion in Neurobiology,* 4 157–165.

120 *If these cells are damaged in an adult* Damasio, A, Yamada, T, Damasio, H, Corbett, J, McKee, J (1980) Central achromatopsia: behavioral, anatomic, and physiologic aspects. *Neurology,* 30: 1064–1071.

121 *Experimental evidence shows that multiple cells from V1* Felleman DJ, Xiao Y, McClendon E (1997) Modular organization of occipito-temporal pathways: cortical connections between visual area 4 and visual area 2 and posterior inferotemporal ventral area in macaque monkeys. *Journal of Neuroscience,* 17(9): 3185–200.

123 *For instance, at least one study has demonstrated that* Annis, RC, Frost, B (1973) Human visual ecology and orientation anisotropies in acuity. *Science,* 182: 729–731.

126 *In the parlance of evolutionary biology* Ryan, MJ (1998) Sexual selection, receiver biases, and the evolution of sex differences. *Science,* 281: 1999–2003.

126 *There are compelling examples where a physical* For example, see Ryan, MJ (1998) Sexual selection, receiver biases, and the evolution of sex differences. *Science,* 281: 1999–2003.

127 *Rather, stimulation of the frog's* Ibid.

127 *sexual selection is often used as a theoretical* For example, see Ridley, M (2003) *The Red Queen: Sex and Evolution of Human Nature.* HarperPerennial, New York; Miller, G (2001) *The Mating Mind: How Sexual Choice Shaped the Evolution of Human Nature.* Anchor Books, New York; Buss, D (1995) *The Evolution of Desire: Strategies of Human Mating.* Basic Books, New York.

127 *The pioneering biologist Amotz* Zahavi, A, Zahavi, A (1997) *The Handicap Principal: A Missing Piece of Darwin's Puzzle.* Oxford University Press, Oxford, England.

128 *However, many of these "improvements"*   Miller, G (2001) *The Mating Mind: How Sexual Choice Shaped the Evolution of Human Nature.* Anchor Books, New York.

128 *Although there is notable cultural*   For an entertaining review, see Etcoff, N (2000) *Survival of the Prettiest.* Anchor Books, New York.

129 *Babies who have an innate fondness for faces*   Thornhill, R, Gangestad, SW (1993) Human facial beauty: averageness, symmetry, and parasite resistance. *Human Nature,* 4: 237–269.

9 Pleasure from Proportion and Symmetry

138 *The English geneticist Angus*   Bateman, A J 1948. Intra-sexual selection in Drosophila. *Heredity* 2: 349–368.

138 *Women produce approximately four hundred*   Williams, GC (1975) *Sex and Evolution.* Princeton University Press, Princeton, NJ.

139 *We can find some help here in the work*   Hamilton, WD, Zuk, M (1982) Heritable true fitness and bright birds: a role for parasites? *Science* 218: 384–387.

140 *Either way, it has been shown*   The best book to find examples of this is the first, Darwin, C, *The Descent of Man.*

140 *Indeed, the handicap principle*   Zahavi, A (1975) Mate selection: A selection for a handicap. *Journal of Theoretical Biology,* 53: 205–214.

142 *Plato and Plotinus wrote extensively*   Hofstader, A, Kuhns, R (eds.) (1964) *Philosophies of Art and Beauty: Selected Readings in Aesthetics from Plato to Heidegger.* University of Chicago Press, Chicago, IL.

144 *Body shape is driven by the distribution*   Singh, D (2002) Female mate value at a glance: Relationship of waist-to-hip ratio to health, fecundity, and attractiveness. *Neuroendocrinology Letters,* 23(4); 81–91.

144 *Increased testosterone in postpubertal boys*   Bjorntorp, P (1991) Adipose tissue distribution and function. *International Journal of Obesity,* 15: 67–81.

144 *Healthy premenopausal women typically*   Singh, D (1993a) Adaptive significance of female physical attractiveness: role of waist-to-hip ratio. *Journal of Personality and Social Psychology,* 65: 292–307.

145 *In the early 1990s, psychologist Devendra*   Ibid.

145 *The drawings seen as least attractive*   Ibid. Singh, D. (1993b) Body shape and women's attractiveness: The critical role of waist-to-hip ratio. *Human Nature,* 4: 297–321.

146 *For these variables, positive rankings*   Singh, D, Luis, S (1995) Ethnic and gender consensus for the effects of waist-to-hip ratio on judgement of women's attractiveness. *Human Nature,* 6: 51–65; Singh, D (2004) Mating strategies of young women: role of physical attractiveness. *The Journal of Sex Research,* 41: 43–54.

146 *Granted, but this is unlikely*   Ibid., Singh, Luis; ibid., Singh, D.

146 *The classic beauties Marilyn Monroe*   Etcoff, N (1999) *Survival of the Prettiest: The Science of Beauty.* Anchor Books, New York.

147 *Women with WHR lower than .8 have a significantly*   Singh, D (1993a); Singh, D (2002).

147 *Women with WHRs below .8 are also significantly more likely*   Ibid., Singh, D; Ibid., Singh, D.

148 *One important marker of phenotypic*   Scheib, JE, Gangestad, SW, Thornhill, R (1999) Facial attractiveness, symmetry and cues of good genes. *Proceedings of the Royal Society of London B,* 266: 1913–1917; Gangestad, SW, Thornhill, R., Yeo, RA. (1994) Facial attractiveness, developmental stability, and fluctuating asymmetry. *Ethology and Sociobiology,* 15: 73–85; Moller, AP, Thornhill, R (1998) Bilateral symmetry and sexual selection: A meta-analysis. *American Naturalist,* 151: 174–192.

149 *Considering this, it is thought that fluctuating asymmetry*   Moller, AP, Swaddle, JP (1997) *Developmental Stability and Evolution.* Oxford University Press, Oxford, England.

149 *Increases are associated with decrements in biological fitness*   For an excellent review, see Thornhill, R, Moller, AP (1997) Developmental stability, disease and medicine. *Biological Reviews* 72: 497–548.

149 *Insofar as fluctuating asymmetry has been found to be partly heritable*   Ibid.; Moller, AP (1990) Fluctuating asymmetry in male sexual ornaments may reliably reveal quality. *Animal Behavior,* 40: 1185–1187; Watson, PJ, Thornhill, R (1994) Fluctuating asymmetry and sexual selection. *Trends in Ecology and Evolution,* 9: 21–25.

149 *Indeed, in the majority of species tested*   Moller, AP, Thornhill, R (1998) Bilateral symmetry and sexual selection: a meta-analysis. *American Naturalist,* 151: 174–192.

149 *In a large-scale review of sixty-five*   Ibid.

150 *A second important finding was that in most species*   In general, secondary sexual traits exhibit greater variability in asymmetry than other traits. Moller, AP, Pomiankowski, A (1993) Fluctuating asymmetry and sexual selection. *Genetica,* 89: 267–279.

150 *For instance, in the dozen or so studies of humans*   Moller, AP, Thornhill, R (1998) Bilateral symmetry and sexual selection: a meta-analysis. *American Naturalist,* 151: 174–192.

153 *They will also visually track a line drawing of a face*   Maurer, D, Young, R (1983) Newborns' following of natural and distorted arrangements of facial features. *Infant Behavior and Development,* 6: 127–131.

153 *Right out of the womb, babies have a preference*   Bushnell, IWR, Sai, F, et al. (1989) Neonatal recognition of mother's face. *British Journal of Developmental Psychology,* 7: 3–15.

153 *By day three, infants can mimic certain facial*   Meltzoff, AN, Moore, MK (1977) Imitation of facial and manual gestures by human neonates. *Science,* 198: 75–78.

153 *Add a few months and infants develop*   Barrera, ME, Maurer, D (1981) Discrimination of stranger by the three-month-old. *Child Development,* 52: 558–563.

153 *and detect different emotional expressions*   Schwartz, GM, Izard, CE, et al. (1985) The 5-month-old's ability to discriminate facial expressions of emotion. *Infant Behavior and Development,* 8: 65–77.

153 *The "face as a kin recognition device" theory*   See one of my favorite books for an excellent review: Konner, M (2001) *The Tangled Wing: Biological Constraints on the Human Spirit*. Henry Holt and Company, New York.

153 *For instance, newborns have a preference for stimuli*   Bornstein, MH, Ferdinandsen, K, et al. (1981) Perception of symmetry in infancy. *Developmental Psychology*, 17: 82–86.

154 *They also prefer objects that are smooth*   Langlois, JH, Roggman, RJ, et al. (1987) Infant preferences for attractive faces: Rudiments of a stereotype? *Developmental Psychology*, 23: 363–369; McCall, RB, Melsom, WH (1970) Complexity, contour, and area as determinants of attention in infants. *Developmental Psychology*, 3: 343–349.

154 *In a series of compelling studies, psychologist Judith*   Ibid., Langlois, Roggman, et al.

155 *Langlois found that infants spent significantly more*   Ibid.

155 *This result has since been replicated and extended*   Langlois, JH, Ritter, JM, et al. (1991) Facial diversity and infant preferences for attractive faces. *Developmental Psychology*, 27: 79–84.

155 *Cross-cultural studies have been done*   See, for example, Jones, D, Hill, K (1993) Criteria for facial attractiveness in five populations. *Human Nature*, 4: 271–295; Wagatsuma, E, Kleinke, CL (1979) Ratings of facial beauty by Korean-Americans and Caucasian females. *Journal of Social Psychology*, 109: 299–300.

156 *Studies by evolutionary biologists and psychologists*   Symons, D (1979) *The Evolution of Human Sexuality*. Oxford University Press, Oxford, England; Langlois, JH, Roggman, LA, et al. (1994) What is average and what is not average about attractive faces? *Psychological Science*, 5: 214–219; Rhodes, G, Tremewan, T (1996) Averageness, exaggeration, and facial attractiveness. *Psychological Science*, 7: 105–110; Grammer, K, Thornhill, R (1994) Human (homo sapiens) facial attractiveness and sexual selection: the role of symmetry and averageness. *Journal of Comparative Psychology*, 108: 233–242.

156 *Randy Thornhill and psychologist Steve Gangestad*   Thornhill, R, Gangestad, SW (1993) Human facial beauty: averageness, symmetry and parasite resistance. *Human Nature*, 4: 237–269.

156 *Other studies have found that people*   Cunningham, MR, Barbee, AP, et al. (1990) What do women want? facial metric assessment of multiple motives in the perception of male physical attractiveness. *Journal of Personality and Social Psychology*, 59: 61–72.

156 *These changes, in turn, raise the metabolic*   Grossman, CJ (1985) Interactions between the gonadal steroids and the immune system. *Science*, 227: 257–261.

157 *In addition to averageness and attention to secondary*   For nice reviews, see Rhodes, G (2006) The evolutionary psychology of facial beauty. *Annual Reviews of Psychology*, 57: 199–226; Thornhill, R, Gangestad, SW (1999) Facial attractiveness. *Trends in Cognitive Sciences*, 3: 452–460.

157 *In a diverse range of species tested*   For a review, see Thornhill, R, Moller, AP (1997) Developmental stability, disease and medicine. *Biological Reviews*, 72: 497–548.

157 *Population biologist John Manning* Manning, JT, Scutt, D, et al. (1998) Developmental stability, ejaculate size, and sperm quality in men. *Evolution and Human Behavior,* 19: 273–282.

157 *They found that men with greater body asymmetry* Ibid.

157 *In women, breast asymmetry* Manning, JT, Scutt, D, et al. (1997) Breast asymmetry and phenotypic quality in women. *Evolution and Human Behavior,* 18: 1–13.

157 *and to the probability of marriage* Moller, AP, Soler, M, et al. (1995) Breast asymmetry, sexual selection and human reproductive success. *Ethology and Sociobiology,* 16: 207–219.

158 *Gangestad and Thornhill gave a series of health questionnaires to* Gangestad, SW, Thornhill, R (1997a) Human sexual selection and developmental stability. In *Evolutionary Social Psychology,* ed. Simpson, JA, and Kenrick, DT. Lawrence Erlbaum Associates, Hillsdale, NJ, 169–195.

158 *The researchers found a significant negative correlation* Ibid.

158 *In a study of 101 college students* Shakleford, TK, Larsen, RJ (1997) Facial asymmetry as an indicator of psychological, emotional, and physiological distress. *Journal of Personality and Social Psychology,* 72: 456–466.

159 *They were also more likely to complain of depression* Ibid.

159 *Consistent with this observation* Markow, TA, Wandler, K (1986) Fluctuating dermatoglyphic asymmetry and the genetics of liability to schizophrenia. *Psychiatry Research,* 19: 323–328; Mellor, CS (1992) Dermatoglyphic evidence of fluctuating asymmetry in schizophrenia. *British Journal of Psychiatry,* 160: 467–472; Durfee, KE (1974) Crooked ears and bad boy syndrome: Asymmetry as an indicator of minimal brain dysfunction. *Bulletin of the Menninger Clinic,* 38: 305–316.

159 *Independent judgments by external observers* Shakleford, Larsen (1997).

159 *This study confirms other reports that facial* See, for example, Grammer, K, Thornhill, R (1994) Human facial attractiveness and sexual selection: the role of symmetry and averageness. *Journal of Comparative Psychology,* 108: 233–242.

161 *Such redundancy may facilitate the increased recognition* Bornstein, MH, Ferdinandsen, K, et al. (1981) Perception of symmetry in infancy. *Developmental Psychology,* 17: 82–86.

161 *Indeed, the data show the latter to be true, since newborns* See, for example, McCall, RB, Melsom, WH (1970) Complexity, contour, and area as determinants of attention in infants. *Developmental Psychology,* 3: 343–349.

161 *Newborns tend to look longer* See, for example, McCall, RB, Melsom, WH (1970) Complexity, contour, and area as determinants of attention in infants. *Developmental Psychology,* 3: 343–349; Bronson, GW (1982) *The Scanning Patterns of Human Infants: Implications for Visual Learning.* Ablex Publishing, Norwood, NJ; Slater, AM (1998) *Perceptual Development: Visual, Auditory and Speech Perception in Infants.* Psychology Press, East Sussex, UK.

161 *Newborns also prefer line drawings of faces* Maurer, D, Young, R (1983) Newborns' following of natural and distorted arrangements of facial features. *Infant Behavior and Development,* 6: 127–131.

162 *As we have just reviewed, body and facial symmetry*   Manning, JT, Scutt, D, et al.
(1998) Developmental stability, ejaculate size, and sperm quality in men. *Evolution and Human Behavior,* 19: 273–282; Manning, Scutt, et al. (1997); Moller, Soler, et al. (1995).

162 *Hence there is evidence that individuals with greater symmetry*   Shakleford, Larsen (1997); Grammer, Thornhill (1994).

163 *In a landmark study, evolutionary psychologist*   For a comprehensive review, see Buss, DM (1994) *The Evolution of Desire.* Basic Books, New York.

163 *For instance, in comparison to men with high asymmetry*   Gangestad, SW, Thronhill, R, et al. (1994) Facial attractiveness, developmental stability and fluctuating asymmetry. *Ethology and Sociobiology,* 15: 73–85; Thornhill, R, Gangestad, SW (1994) Human fluctuating asymmetry and sexual behavior. *Psychological Science,* 5: 297–302; Gangestad, Thornhill (1997a); Gangestad, SW, Thorhill, R (1997b) The evolutionary psychology of extra-pair sex: the role of fluctuating asymmetry. *Ethology and Sociobiology,* 18: 69–88; Thornhill, R, Gangestad, SW, et al. (1995) Human female orgasm and mate fluctuating asymmetry. *Animal Behavior,* 50: 1601–1615.

163 *In a recent study, biologist Craig Roberts*   Roberts, SC, Little, AC, et al. (2005) MHC-heterozygosity and human facial attractiveness. *Evolution and Human Behavior,* 26: 213–226.

164 *Men with greater heterozygosity*   Ibid.

164 *As we saw earlier, waist-to-hip*   Singh, D (1993a); Singh, D (1993b); Singh, D, Luis, S (1995); Singh, D (2004).

164 *In other experiments, Devendra Singh*   Singh, D (1995) Female health, attractiveness, and desirability for relationships: role of breast asymmetry and waist-to-hip ratio. *Ethology and Sociobiology,* 16: 465–481.

164 *Whereas most body parts exhibit fluctuating*   Moller, AP, Soler, M, et al. (1995) Breast asymmetry, sexual selection and human reproductive success. *Ethology and Sociobiology,* 16: 207–219; Manning, JT, Scutt, D, et al. (1996) Asymmetry and menstrual cycle in women. *Ethology and Sociobiology,* 17: 129–143.

165 *For instance, similar to newborns*   Palmer, SE, Hemenway, K (1978) Orientation and symmetry: effects of multiple, near, and rotational symmetries. *Journal of Experimental Psychology: Human Perception and Performance,* 4: 691–702; Royer, F (1981) Detection of symmetry. *Journal of Experimental Psychology: Human Perception and Performance,* 7: 1186–1210.

165 *Moreover, symmetric objects and patterns*   Humphrey, D (1997) Preferences in symmetries and symmetries in drawings: asymmetries between ages and sexes. *Empirical Studies of the Arts,* 15: 41–60; Berlyne, DE (1974) *Studies in the New Experimental Aesthetics: Steps Toward an Objective Psychology of Aesthetic Appreciation.* Hemisphere Company, Washington DC; Reber, R, Schwarz, N (2006) Perceptual fluency, preference, and evolution. *Polish Psychological Bulletin,* 37:16–22; Gombridge, EH (1984) *The Sense of Order: A Study in the Psychology of Decorative Art.* Phaidon, London.

165 *Indeed, there is widespread use of symmetric designs*   Ibid. Gombridge; McManus, C (2002) *Right Hand, Left Hand: The Origins of Asymmetry in Brains, Bodies, Atoms, and Cultures.* Harvard University Press, Cambridge, MA.

165 *The subjects were asked to "choose the design*   Cardenas, RA, Harris, LJ (2006) Symmetrical decorations enhance the attractiveness of faces and abstract designs. *Evolution and Human Behavior,* 27: 1–18.

166 *This suggests that symmetry is preferred in nonbiological*   Ibid.

167 *Interestingly, the application of an asymmetric design to a symmetric face*   Ibid.

10 Pleasure from Repetition and Rhythm

170 *After only a few months, the scientists found*   Chang, EF, Merzenich, MM (2003) Environmental noise retards auditory cortical development. *Science,* 300: 498–502.

171 *Most interesting of all, when the noise-reared rats*   Ibid.

171 *For instance, it is commonly known that older infants*   Dissanayake, E (1992) *Homoaestheticus.* Free Press, New York.

171 *Various religious groups practice meditation*   See, for example, Gass, R, Brehony, KA (2000) *Chanting: Discovering Spirit in Sound.* Broadway, New York; Hahn, TN (2006) *Chanting from the Heart: Buddhist Ceremonies and Daily Practices.* Parallax Press, Berkeley, CA; Crummet, M (1993) *Sun Dance.* Falcon Press Publishing, San Ramon, CA.

171 *Babies as young as four months old*   Schellenberg, EG, Trehub, SE (1996) Natural musical intervals: evidence from infant listeners. *Psychological Science,* 7: 272–277.

172 *As an example, let us consider music*   Miller, GF (2000) *The Mating Mind.* Doubleday, New York; Levitan, DJ (2006) *This Is Your Brain on Music.* Dutton, New York; Mithen, S (2006) *The Singing Neanderthals: The Origins of Music, Language, Mind, and Body.* Harvard University Press, Cambridge, MA.

172 *Dawkins argued that the classic idea*   Dawkins, R (1982) *The Extended Phenotype: The Long Reach of the Gene.* Oxford University Press, Oxford, England.

173 *Rather, he suggests that in addition to its potential importance*   Mithen, S (2006).

173 *Following Darwin's lead, he suggests that music*   Miller, GF (2000) Evolution of human music through sexual selection. In Wallin, NL, Nerker, B, Brown, S (eds.), *The Origins of Music.* MIT Press, Cambridge, MA. Also see Mithen, S (2006), who suggests that both natural and sexual selection have driven the evolution of human music production and perception.

174 *He notes that from this sample, "males*   Miller, GF (2000).

174 *It is difficult to imagine the hunter-gatherer equivalent*   This example was also considered by Miller, GF (2000).

174 *But as Dawkins and others have observed, examples abound*   Ibid. Dawkins, R (1982); and see Hauser, M (1996) *The Evolution of Communication.* MIT Press, Cambridge, MA.

175 *Studies in rodents and primates have found*   For example, see Chang, EF, Merzenich, MM (2003).

175 *Clearly, music has effects on social communication*   Mithen, S (2006).

176 *That is, they exhibit learning similar*   Mache, FB (2000) The necessity and problem with a universal musicality. In Wallin, NL, Nerker, B, Brown, S (eds.), *The
Origins of Music.* MIT Press, Cambridge, MA.

176 *There is now evidence that females*   Catchpole, CK, Slater, PJB (1995) *Bird
Song: Biological Themes and Variations.* Cambridge University Press, Cambridge,
England.

176 *Many species of birds also sing songs*   Mache, FB (2000); Catchpole, CK, Slater,
PJB (1995).

178 *Likewise, virtuosic performance of instrumental*   Miller, GF (2000).

179 *Indeed, experiments have demonstrated that when speaking*   Bolinger, D (1986)
Intonation and Its Parts: Melody in Spoken English. Stanford University Press,
Palo Alto, CA.

11 Homo Addictus

182 *Rats that are made sick by ingesting tainted*   Reilly, S, Bornovalova, MA (2005)
Conditioned taste aversion and amygdala lesions in the rat: A critical review.
Neuroscience and Biobehavioral Reviews, 29(7): 1067–1088; Welzl, H, D'Adamo,
P, Lipp, HP (2001) Conditioned taste aversion as a learning and memory
paradigm. *Behavioral Brain Science,* 125(1–2): 205–213; Yamamoto, T, Shimura,
T, Sako, N, et al. (1994) Neural substrates for conditioned taste aversion in the
rat. *Behavioral Brain Research,* 65(2): 123–137.

188 *Negative feelings occur with the presence of fitness decrements*   Panksepp,
J, Knutson, B, Burgdorf, J (2002) The role of brain emotional systems in
addiction: a neuro-evolutionary perspective and new "self-report" animal
model. *Addiction,* 97: 459–469.

188 *Many anthropologists have pointed out*   For example, see Sullivan, RJ, Hagan, EH
(2002) Psychotropic substance seeking: evolutionary pathology or adaptation.
Addiction, 97: 389–400; Dudley, R (2002) Fermenting fruit and the historical
ecology of ethanol ingestion: is alcoholism in modern humans an evolutionary hangover? *Addiction,* 97: 381–388.

189 *For instance,* Areca catechu, *commonly known*   Glover, IC (1977) Prehistoric
plant remains from Southeast Asia with special reference to rice. In Taddei, M
(ed.), *South Asian Archaeology.* Instituto Universitario Orientale, Naples, 7–37.

189 *There is also evidence that nicotine*   Watson, P (1983) *This Precious Foliage: A Study
of the Aboriginal Psychoactive Drug Pituri.* University of Sydney, Sydney, Australia.

189 *For many psychoactive substances*   Sullivan, RJ, Hagan, EH, 2002.

189 *Indeed, the anthropoid diet has been predominantly*   Dudley, R, 2002.

189 *Temperate-zone fruit sources*   McKenzie, JA, McKechnie, W (1979) A comparative study of resource utilization in natural populations of *Drosophila
Malanogaster* and *D. simulans. Oecologia,* 40: 299–309.

189 *Comparative studies have found that as most temperate*   Brady, CJ (1987) Fruit
ripening. *Annual Review of Plant Physiology,* 38: 155–178.

189 *Some anthropologists have suggested that ethanol*   Dudley, R (2002).

190 *More than 60 percent of Americans have tried an illicit*   Johnston, LD, O'Malley, PM, Bachman, JG (2001) *Monitoring the Future: National Survey Results on Drug Use, 1975–2000.* Volume II. *College Students and Adults Aged 19–40.* Bethesda, MD, National Institute on Drug Abuse, MIH Publication 01-4925.

190 *For instance, recent studies have found that even for a highly*   Wagner, FA, Anthony, JC (2002) From first drug use to drug dependence: Developmental periods of risk for dependence upon marijuana, cocaine, and alcohol. *Neuropsychopharmacology,* 26: 479–488.

195 *possessing an iron frame*   Harlow, JM (1868) Recovery from the passage of an iron bar through the head. *Publications of the Massachusetts Medical Society,* 2: 329–346; Harlow, JM (1848–1849) Passage of an iron rod through the head. *Boston Medical and Surgical Journal,* 39: 389.

195 *This exacting and decisive nature*   Ibid.

195 *"to please the fancy of the owner"*   Bigelo, HJ (1850) Dr. Harlow's case of recovery from the passage of an iron bar through the head. *American Journal of the Medical Sciences,* 19: 13–22.

196 *"Gage is no longer Gage"*   Ibid.

197 *"possessed a well-balanced mind*   Ibid. Harlow, JM (1868).

197 *A child in his intellectual capacity and manifestations*   Ibid.

198 *There is emerging evidence that chronic exposure*   Jentsch, JD, Taylor, JR (1999) Impulsivity resulting from frontostriatal dysfunction in drug abuse: implications for the control of behavior by reward-related stimuli. *Psychopharmacology,* 146: 373–390; Volkow, ND, Hitzemann, R, Wang, GJ, et al. (1992) Long-term frontal brain metabolic changes in chronic cocaine abusers. *Synapse,* 11: 184–190.

198 *Consistent with these findings, addicts*   For example, Jentsch and Taylor (1999); Bechara, A, Damasio, H (2002) Decision-making and addiction. Part I. Impaired activation of somatic states in substance dependent individuals when pondering decisions with negative future consequences. *Neuropsychologia,* 40: 1675–1689.

198 *It is very likely that a loss of inhibitory*   Robinson, TE, Berridge, KC (2003) Addiction. *Annual Review of Psychology,* 54: 25–53.

198 *This view is rooted in fairly recent findings*   Robinson, TE, Berridge, KC (1993) The neural basis of drug craving: An incentive-sensitization theory of addiction. *Brain Research Reviews,* 18: 157–198; Robinson, TE, Berridge, KC (2000) The psychology and neurobiology of addiction: an incentive-sensitization view. *Addiction,* 95: S91–S117; Robinson, TE, Berridge, KC (2003).

199 *A major point, however, is that these animals*   See, for example, Pecina, S, Cagniard, B, Berridge, KC, et al. (2003) Hyperdopaminergic mutant mice have higher "wanting" but not "liking" for sweet rewards. *Journal of Neuroscience,* 23: 9395–9402.

199 *Injection of chemicals that boost*   Parker, LA, Maier, S, Rennie, M, et al. (1992) Morphine- and naltrexone-induced modification of palatability: Analysis by the taste reactivity test. *Behavioral Neuroscience,* 106: 999–1010.

200 *Working in Berridge's laboratory* For an excellent review see Pecina, S, Smith, KS, Berridge, KC (2006) Hedonic hot spots in the brain. *Neuroscientist,* 12(6): 500–511.

201 *Work in the late 1980s* Berridge, KC (1988) Brainstem systems mediate the enhancement of palatability by chlordiazepoxide. *Brain Research,* 447: 262–268.

203 *Chronic stress and the associated activation* For a review see Wallenstein, GV (2002) *Mind Stress, and Emotions: The New Science of Mood.* Commonwealth Press, Boston.

203 *For instance, in a landmark study, ethologist Dee Higley* Higley, JD, Hasert, MF, Suomi, SJ, et al. (1991) Nonhuman primate model of alcohol abuse: Effects of early experience, personality, and stress on alcohol consumption. *Proceedings of the National Academy of Sciences,* 88: 7261–7265.

204 *Another potential therapeutic target might be the opioid* Panksepp, J (1998) *Affective Neuroscience: The Foundations of Human and Animal Emotions.* Oxford University Press, New York.

204 *Clonidine, an alpha-1 noradrenergic* Gold, MS (1993) Opiate addiction and the locus coeruleus: the clinical utility of clonidine, natrexone, methadone, and buprenophine. *Psychiatric Clinics of North America,* 16: 61–73.

12 Parsing Pleasure

209 *Indeed, becoming addicted to either smoking or drinking alcohol* Pomerleau, CS, et al. (2004) Relationship between early experiences with tobacco and early experiences with alcohol. *Addictive Behaviors,* 29(6): 1245–51.

210 *This was the first broadly read account of how the advertising community* Packard, V (1957) *The Hidden Persuaders.* Pocket Books, New York.

211 *Our innate preference for these features has been studied by marketing* See, for example, Bloch, PH (1995) Seeking the ideal form: Product design and consumer response. *Journal of Marketing,* 59(3): 16–29.

Index